Lecture Notes in Computer Science 8466

Commenced Publication in 1973
Founding and Former Series Editors:
Gerhard Goos, Juris Hartmanis, and Jan van Leeuwen

T0236429

Reneta P. Barneva   Valentin E. Brimkov
Josef Šlapal (Eds.)

Combinatorial Image Analysis

16th International Workshop, IWCIA 2014
Brno, Czech Republic, May 28-30, 2014
Proceedings

 Springer

Volume Editors

Reneta P. Barneva
State University of New York at Fredonia
Department of Computer and Information Sciences
280 Central Ave., Fredonia, NY 14063, USA
E-mail: barneva@fredonia.edu

Valentin E. Brimkov
SUNY Buffalo State College
Mathematics Department
1300 Elmwood Ave., Buffalo, NY 14222, USA
E-mail: brimkove@buffalostate.edu

Josef Šlapal
Brno University of Technology
Institute of Mathematics
616 69 Brno, Czech Republic
E-mail: slapal@fme.vutbr.cz

ISSN 0302-9743   e-ISSN 1611-3349
ISBN 978-3-319-07147-3   e-ISBN 978-3-319-07148-0
DOI 10.1007/978-3-319-07148-0
Springer Cham Heidelberg New York Dordrecht London

Library of Congress Control Number: 2014938394

LNCS Sublibrary: SL 6 – Image Processing, Computer Vision, Pattern Recognition, and Graphics

Typesetting: Camera-ready by author, data conversion by Scientific Publishing Services, Chennai, India

Printed on acid-free paper

Springer is part of Springer Science+Business Media (www.springer.com)

Preface

The present volume includes the articles presented at the 16th International Workshop on Combinatorial Image Analysis, IWCIA 2014, which was held in Brno (Czech Republic), May 28–30, 2014. The 15 previous meetings were held in Paris (France) 1991, Ube (Japan) 1992, Washington DC (USA) 1994, Lyon (France) 1995, Hiroshima (Japan) 1997, Madras (India) 1999, Caen (France) 2000, Philadelphia (USA) 2001, Palermo (Italy) 2003, Auckland (New Zealand) 2004, Berlin (Germany) 2006, Buffalo, NY (USA) 2008, Playa del Carmen (Mexico) 2009, Madrid (Spain) 2011, and Austin, TX (USA) 2012.

Image analysis provides theoretical foundations and methods for solving problems from various areas of human practice. In combinatorial image analysis the models are discrete and integer arithmetic is used. The developed algorithms rely on combinatorial properties of digital images and often outperform algorithms based on continuous models, float arithmetic, and rounding.

The submission and review process of the workshop was carried out through the OpenConf conference management system. The reviewing process was quite rigorous, including double-blind reviews by at least three highly qualified members of the international Program Committee. As a result, 20 papers authored by researchers from 11 different countries were accepted for presentation at the workshop and for inclusion in this volume. The most important selection criterion was the overall score received. Other criteria included relevance to the workshop scope, correctness, novelty, clarity, and presentation quality. We believe that as a result only papers of very high quality were accepted for publication in the workshop proceedings and for presentation at IWCIA 2014.

The program of the workshop included presentations of contributed papers and keynote talks given by three distinguished scholars. An opening talk given by Gabor Herman (Graduate Center, City University of New York) presented the recently developed (by him and others) superiorization methodology. The methodology provides an optimization criterion that helps to distinguish the "better" among a large set of constraints-compatible solutions, which are typically available when solving constrained optimization problems related to scientific, engineering, and medical applications of image analysis.

Valentin Brimkov (Buffalo State College, State University of New York) discussed parallel computation techniques for two-dimensional combinatorial pattern matching problems, the latter being relevant to pattern recognition, low level image processing, computer vision, and multimedia. He illustrated some key ideas by his recent results on detection of two-dimensional repetitions in a two-dimensional array.

Shadia Rifai Habbal (Institute for Astronomy, University of Hawaii) presented recently developed (by her and others) new methods that yield artifact-free images and uncover details that are hidden in the original unprocessed images. This has led to the discovery of new features that are essential for exploring the dynamics and thermodynamics of structures in the solar corona.

Extended abstracts of the keynote talks are included at the beginning of this volume.

The contributed papers included in the volume are grouped into three sections. Seven of them are devoted to problems of discrete geometry and topology and their use in imaging science; six papers present new results on image representation, segmentation, grouping, and reconstruction; and seven works present applications in medical image processing and other fields.

We believe that the attendees have benefited from the scientific program. We also hope that many of the papers can be of interest to a broader audience, including researchers in the areas of computer graphics and computer vision.

In addition to the main track of the workshop, a Special Track for Applications took place at IWCIA 2014. It provided researchers with the opportunity to present their last developments and implementations.

We would like to thank everyone who contributed to the success of IWCIA 2014. First of all, the workshop chair is indebted to IWCIA's Steering Committee for endorsing the candidacy of Brno for the 16th edition of the workshop, and to the keynote speakers Gabor Herman, Valentin Brimkov, and Shadia Rifai Habbal for their excellent talks and overall contribution to the workshop.

Sincere thanks go to the members of IWCIA's Program Committee whose timely and high-quality reviews were essential in establishing a strong workshop program. We thank all researchers who submitted works to IWCIA 2014. Thanks to their contributions we succeeded in having a technical program of high scientific quality. We wish to thank the participants and all who contributed to making this workshop an enjoyable and fruitful scientific event. We also thank the organizers for the excellent conditions and the pleasant time we all had in Brno. Finally, we express our gratitude to Springer and especially to Alfred Hofmann and Anna Kramer, for the efficient and kind cooperation in the timely production of this book.

May 2014   Reneta P. Barneva
                                                          Valentin E. Brimkov
                                                              Josef Šlapal

Organization

IWCIA 2014 was held at Brno University of Technology, Czech Republic, May 28–30, 2014.

General Chair

Josef Šlapal — Brno University of Technology, Czech Republic

Program and Publication Chair

Reneta P. Barneva — SUNY Fredonia, USA

Steering Committee

Valentin E. Brimkov — SUNY Buffalo State College, USA
Gabor T. Herman — CUNY Graduate Center, USA
Kostadin Koroutchev — Universidad Autonoma de Madrid, Spain
Petra Wiederhold — CINVESTAV-IPN, Mexico

Invited Speakers

Valentin E. Brimkov — SUNY Buffalo State College, USA
Gabor T. Herman — CUNY Graduate Center, USA
Shadia Rifai Habbal — Institute for Astronomy, University of Hawaii, USA

Program Committee

Akira Asano — Kansai University, Japan
Péter Balázs — University of Szeged, Hungary
Reneta P. Barneva — SUNY Fredonia, USA
George Bebis — University of Nevada at Reno, USA
Bhargab B. Bhattacharya — Indian Statistical Institute, India
Jean-Marc Chassery — University of Grenoble, France
Marco Cristani — University of Verona, Italy
Guillaume Damiand — LIRIS-CNRS, Université de Lyon, France

Eduardo Destefanis	Universidad Tecnologica Nacional Córdoba, Argentina
Chiou-Shann Fuh	CSIE National Taiwan University, Taiwan
Rocío González Díaz	University of Seville, Spain
Atsushi Imiya	IMIT, Chiba University, Japan
María José Jiménez	University of Seville, Spain
Ramakrishna Kakarala	NTU, Singapore
Walter G. Kropatsch	Vienna University of Technology, Austria
Joakim Lindblad	University of Novi Sad, Serbia
Hongbing Lu	Fourth Military Medical University, China
Pavel Matula	Masaryk University, Czech Republic
Petr Matula	Masaryk University, Czech Republic
Benedek Nagy	University of Debrecen, Hungary
Akira Nakamura	Hiroshima University, Japan
Kalman Palagyi	University of Szeged, Hungary
Petra Perner	Institute of Computer Vision and Applied Computer Sciences, Germany
Hemerson Pistori	Dom Bosco Catholic University, Brazil
Ioannis Pitas	Aristotle University of Thessaloniki, Greece
Konrad Polthier	Freie Universität Berlin, Germany
Md. Atiqur Rahman Ahad	University of Dhaka, Bangladesh
Xavier Roca Marvà	UAB, Spain
Arun Ross	West Virginia University, USA
Angel Sappa	Computer Vision Center, Spain
Henrik Schulz	Helmholtz-Zentrum Dresden-Rossendorf, Germany
Nikolay M. Sirakov	Texas A&M University, USA
Rani Siromoney	Madras Christian College, India
Alberto Soria	CINVESTAV, Mexico
K.G. Subramanian	Universiti Sains Malaysia, Malaysia
Akihiro Sugimoto	National Institute of Informatics, Japan
Mohamed Tajine	University Louis Pasteur, Strasbourg, France
Joao Manuel R.S. Tavares	University of Porto, Portugal
Peter Veelaert	Ghent University, Belgium
Young Woon Woo	Dong-Eui University Busan, Korea
Jinhui Xu	SUNY University at Buffalo, USA
Yasushi Yagi	Osaka University, Japan
Richard Zanibbi	Rochester Institute of Technology, USA
Pavel Zemčík	Brno University of Technology, Czech Republic

Organizing Committee

Miloslav Druckmüller	Brno University of Technology, Czech Republic
Hana Druckmüllerová	Brno University of Technology, Czech Republic

Jana Hoderová	Brno University of Technology, Czech Republic
Dalibor Martišek	Brno University of Technology, Czech Republic
Jan Pavlík	Brno University of Technology, Czech Republic
Pavla Sehnalová	Brno University of Technology, Czech Republic
Pavel Štarha	Brno University of Technology, Czech Republic
Michael Szocki	SUNY Fredonia, USA

Sponsoring Institutions

Brno University of Technology, Faculty of Mechanical Engineering

Table of Contents

Image Representation, Segmentation, Grouping, and Reconstruction

Applications in Medical Image Processing and Other Fields

Superiorization for Image Analysis

Gabor T. Herman

Computer Science PhD Program, Graduate Center, City University of New York,
365 Fifth Avenue, New York, NY 10016, U.S.A.
gabortherman@yahoo.com

Abstract. Many scientific, engineering and medical applications of image analysis use constrained optimization, with the constraints arising from the desire to produce a solution that is constraints-compatible. It is typically the case that a large number of solutions would be considered good enough from the point of view of being constraints-compatible. In such a case, an optimization criterion is introduced that helps us to distinguish the "better" constraints-compatible solutions. The superiorization methodology is a recently-developed heuristic approach to constrained optimization. The underlying idea is that in many applications there exist computationally-efficient iterative algorithms that produce solutions that are constraints-compatible. Often the algorithm is perturbation resilient in the sense that, even if certain kinds of changes are made at the end of each iterative step, the algorithm still produces a constraints-compatible solution. This property is exploited in superiorization by using such perturbations to steer the algorithm to a solution that is not only constraints-compatible, but is also desirable according to a specified optimization criterion. The approach is very general, it is applicable to many iterative procedures and optimization criteria. Most importantly, superiorization is a totally automatic procedure that turns an iterative algorithm into its superiorized version. This, and its practical consequences in various application areas, have been investigated for a variety of constrained optimization tasks.

Keywords: Superiorization, Constrained optimization, Perturbation resilient algorithm, Image processing.

1 Overview

Optimization is a tool used in many applications of image analysis in science, engineering and medicine. Examples from medical imaging are: tomographic reconstruction [16, 26, 32, 35, 37, 43], radiation therapy treatment planning [11, 17, 25, 28, 40], image registration [38, 41] and segmentation [20]. Many applications use constrained optimization, with the constraints arising from the desire to produce a solution that is constraints-compatible, in the sense of meeting the requirements of physically or otherwise obtained constraints. For example, in computerized tomography the constraints come from the detector readings of the instrument. In such applications it is typically the case that a large number

R.P. Barneva, V.E. Brimkov, and J. Šlapal (Eds.): IWCIA 2014, LNCS 8466, pp. 1–7, 2014.
© Springer International Publishing Switzerland 2014

of solutions would be considered good enough from the point of view of being constraints-compatible; to a large extent due to the uncertainty as to the exact nature of the constraints (for example, due to noise in the data). In such a case, an optimization criterion is introduced to help us distinguishing the "better" constraints-compatible solutions. Because of their extreme practical importance in many fields, such constrained optimization problems have been extensively studied over many years, as exemplified by the books [1–3, 10, 29, 34].

2   Primary Significant Contribution of Superiorization

The number of combinations of a set of constraints and an optimization criterion that my arise is limitless. In spite of the great deal of knowledge that exists regarding constrained optimization, we keep coming across new combinations for which the existing knowledge is insufficient and a new algorithm has to be developed. It is often the case that the mathematical discovery and implementation of an algorithm for such a new constrained optimization problem is far from trivial, it requires a major investment of time of the researcher and even then success is not guaranteed. On the other hand, it is generally a much easier task to find a computationally-efficient algorithm just for constraints-compatibility (without optimization). Typically such algorithms are iterative, they produce a potentially infinite sequence of images from which we can select the first one that is constraints-compatible at a specified level. Furthermore, often the algorithm is perturbation resilient in the sense that, even if certain kinds of changes are made at the end of each iterative step, the algorithm still produces a constraints-compatible solution. This property is exploited in the superiorization approach by using such perturbations to steer the algorithm to an output that is as constraints-compatible as the output of the original algorithm, but it is superior to it according to the optimization criterion. The aim is to present a totally automatic procedure that turns the iterative algorithm into such a superiorized version. This has been done for a very large class of iterative algorithms and for optimization criteria in general, typical restrictions (such as convexity) on the optimization criterion are not needed. Thus superiorization can be significantly helpful, because it has the potential of saving a lot of time and effort for the researcher when the application of interest gives rise to a new constrained optimization problem. This essential contribution deserves repeating in alternate words: If we have an iterative algorithm for constraints-compatibility (which, in general, is much easier to produce than one for constrained optimization) and if we are given an optimization criterion, then the production of the superiorized version of the algorithm is totally automatic. As an example of this, consider likelihood maximization using the iterative algorithm of expectation maximization (ML-EM) for positron emission tomography (PET) that was first proposed in [36]. For this, likelihood is used as a measure of constraints-compatibility. It was observed that images deteriorate with a large number of iterations, in the sense that they present irregular high amplitude patterns. There have been various approaches to deal with the appearance of these artifacts. One approach is by reformulating the PET reconstruction problem into one of maximum *a posteriori* (MAP) estimation. This was first done in [27], under the assumption that there is a prior Gaussian distribution of PET images in which smoother images have a higher probability and which can be used to provide the optimization criterion. An in-practice-useful contribution of [27] was a MAP-EM algorithm that is very similar in its computational details to the ML-EM algorithm and yet converges to the MAP estimator. However, to guarantee this desirable behavior it was necessary to develop new mathematics; the effort invested in doing that for [27] was significant. In contrast, as discussed in [18] and illustrated in Fig. 1 below, the undesirable behavior of ML-EM can be automatically eliminated by superiorizing it.

3 Secondary Significant Contribution of Superiorization

Constrained optimization problems that arise in applications are often huge. It can then happen that the traditional algorithms for constrained optimization require computational resources that are not available and, even if they are, the length of time needed to produce an acceptable output is too long to be practicable. The computational requirements of the superiorized algorithms are often significantly less than those of the traditional algorithms. See, for example, [7] that reports on a comparison of the projected subgradient method (PSM), which is a standard method of classical optimization, with a superiorization approach, which was found to be over twenty times faster than PSM.

4 Innovative Aspects of Superiorization

The superiorization methodology was first proposed by the author and coworkers a few years ago [4]. It is a heuristic approach to constrained optimization in the sense that the process is not guaranteed to lead to an optimum according to the given criterion; processes that are guaranteed in that sense are usually referred to as exact. Heuristic approaches have been found useful in applications of optimization, because they are often computationally less expensive than their exact counterparts, but nevertheless provide solutions that are appropriate for the application at hand [33, 39, 42]. Superiorization is based on the fact that in many applications there exist computationally-efficient iterative algorithms that produce constraints-compatible solutions for the given constraints. (An example for radiation therapy is in [11].) Often the algorithm is perturbation resilient, which is exploited in the superiorization approach by using perturbations to steer the algorithm to a solution that is not only constraints-compatible, but is also desirable according to an optimization criterion.

The idea of algorithms that interlace steps of two different kinds (in our case, one kind of steps aim at constraints-compatibility and the others aim at improvement of the optimization criterion) has been made use of in many approaches to exact constrained optimization in mind [12, 13, 15, 21, 22, 31, 37]. However, those approaches cannot do what is done by superiorization: the automatic production of a constrained optimization algorithm from an iterative algorithm for constraints-compatibility. While it is true that when applied to any particular constrained optimization problem, the superiorization algorithm may not look much different from one that would have been obtained conventionally, the essential difference is the following: The conventional approach designs an algorithm that depends simultaneously on both the constraints and the optimization criterion (and this may be difficult to do), while the superiorization algorithm takes an iterative algorithm for the constraints (this is often easy to design) and produces from it the superiorized version for the given optimization criterion in an automatic fashion.

The world-view of superiorization is different from that of classical constrained optimization. Both in superiorization and in classical constrained optimization we assume the existence of a domain Ω and an optimization criterion, specified by a function ϕ that has a real-number value for every $\boldsymbol{x}$ in Ω.

In classical optimization it is assumed that there is a constraints set C and the optimization task is to find an $\boldsymbol{x} \in C$ for which $\phi(\boldsymbol{x})$ is minimal. One problem with this is that the constraints may not be consistent and so C could be empty and the optimization task as stated would not have a solution. Another is that iterative methods of classical constrained optimization typically converge to a solution only in the limit and so some stopping rule is applied to terminate the process. The actual output at that time may not be in C and, even if it is in C, it is most unlikely to be a minimizer of ϕ over C.

Both problems are handled in the superiorization approach by replacing the C by a nonnegative real-valued function $\mathcal{P}r$ that indicates of how incompatible a given $\boldsymbol{x} \in \Omega$ is with the constraints. The merit of the output of an algorithm is given by the smallness of the two numbers $\mathcal{P}r(\boldsymbol{x})$ and $\phi(\boldsymbol{x})$. It was shown in [24] that if an iterative algorithm produces an output $\boldsymbol{x}$, then its superiorized version will produce an output $\boldsymbol{x}'$ for which $\mathcal{P}r(\boldsymbol{x}')$ is not larger then $\mathcal{P}r(\boldsymbol{x})$, but (in general) $\phi(\boldsymbol{x}')$ is much smaller than $\phi(\boldsymbol{x})$. It is also the case that the superiorization approach is often more efficacious than classical optimization, in the sense that if the latter produces an actual output $\boldsymbol{x}$, then the superiorization approach will produce an output $\boldsymbol{x}'$ for which $\mathcal{P}r(\boldsymbol{x}')$ is not larger then $\mathcal{P}r(\boldsymbol{x})$, $\phi(\boldsymbol{x}')$ is much smaller than $\phi(\boldsymbol{x})$ and the computer time required to obtain $\boldsymbol{x}'$ is less than that required by the classical approach to obtain $\boldsymbol{x}$.

5   A Brief and Partial Summary of the Superiorization Literature

The superiorization methodology was first proposed in [4]. There perturbation resilience was proved for the general class of string-averaging projection methods [8]. In [23] the advantages of superiorization for image reconstruction from a small number of projections was studied. In [14] perturbation resilience was analyzed for the class of block-iterative (BIP) projection methods [5]. In [6] the methodology was formulated over general problem structures. These papers used various optimization criteria, including minimization of total variation (TV) and

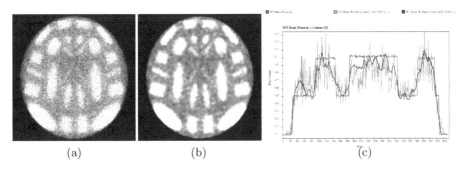

Fig. 1. PET simulation. (a) Reconstruction using EM. (b) Reconstruction using superiorization. (c) Activity along the 323rd column in the phantom (red), in the EM reconstruction (green) and in the superiorization reconstruction (blue).

maximization of entropy; this was extended in [19] to minimization of the ℓ_1-norm of the Haar transform. In [30] two acceleration schemes based on BIP methods were proposed and evaluated. This line of research culminated in [24], which succeeded in providing superiorization with a solid and comprehensive mathematical foundation. Three more recent relevant articles are [7, 9, 18]. An illustration based on the last of these papers is shown in Fig. 1.

Acknowledgment. The work presented here is currently supported by the National Science Foundation (award number DMS-1114901).

References

1. Bazaraa, M., Sherali, H., Shetty, C.: Nonlinear Programming: Theory and Algorithms, 2nd edn. Wiley (1993)
2. Bertsekas, D.: Nonlinear Programming. Athena Scientific (1995)
3. Bertsekas, D., Tsitsiklis, J.: Parallel and Distributed Computation: Numerical Methods. Prentice-Hall (1989)
4. Butnariu, D., Davidi, R., Herman, G.T., Kazantsev, I.: Stable convergence behavior under summable perturbations of a class of projection methods for convex feasibility and optimization problems. IEEE J. Select. Topics Signal Proc. 1, 540–547 (2007)
5. Censor, Y., Chen, W., Combettes, P., Davidi, R., Herman, G.T.: On the effectiveness of projection methods for convex feasibility problems with linear inequality constraints. Comput. Optim. Appl. 51, 1065–1088 (2012)
6. Censor, Y., Davidi, R., Herman, G.T.: Perturbation resilience and superiorization of iterative algorithms. Inverse Probl. 26, 065008 (2010)
7. Censor, Y., Davidi, R., Herman, G.T., Schulte, R.W., Tetruashvili, L.: Projected subgradient minimization versus superiorization. Journal of Optimization Theory and Applications 160, 730–747 (2014)

8. Censor, Y., Elfving, T., Herman, G.T.: Averaging strings of sequential iterations for convex feasibility problems. In: Butnariu, D., Censor, Y., Reich, S. (eds.) Inherently Parallel Algorithms in Feasibility and Optimization and Their Applications, pp. 101–114. Elsevier (2001)
9. Censor, Y., Zaslavski, A.J.: String-averaging projected subgradient methods for constrained minimization. Optimization Methods and Software 29, 658–670 (2014)
10. Censor, Y., Zenios, S.: Parallel Optimization: Theory, Algorithms and Applications. Oxford University Press (1998)
11. Chen, W., Craft, D., Madden, T., Zhang, K., Kooy, H., Herman, G.T.: A fast optimization algorithm for multicriteria intensity modulated proton therapy planning. Med. Phys. 37, 4938–4945 (2010)
12. Combettes, P., Luo, J.: An adaptive level set method for nondifferentiable constrained image recovery. IEEE T. Image Proc. 11, 1295–1304 (2002)
13. Combettes, P., Pesquet, J.C.: Image restoration subject to a total variation constraint. IEEE T. Image Proc. 13, 1213–1222 (2004)
14. Davidi, R., Herman, G.T., Censor, Y.: Perturbation-resilient block-iterative projection methods with application to image reconstruction from projections. Int. T. Oper. Res. 16, 505–524 (2009)
15. Defrise, M., Vanhove, C., Liu, X.: An algorithm for total variation regularization in high-dimensional linear problems. Inverse Probl. 27, 065002 (2011)
16. Dutta, J., Ahn, S., Li, C., Cherry, S., Leahy, R.: Joint L-1 and total variation regularization for fluorescence molecular tomography. Phys. Med. Biol. 57, 1459–1476 (2012)
17. Fredriksson, A., Forsgren, A., Hardemark, B.: Minimax optimization for handling range and setup uncertainties in proton therapy. Med. Phys. 38, 1672–1684 (2011)
18. Garduño, E., Herman, G.T.: Superiorization of the ML-EM algorithm. IEEE Transactions on Nuclear Science 61, 162–172 (2014)
19. Garduño, E., Herman, G.T., Davidi, R.: Reconstruction from a few projections by ℓ_1-minimization of the Haar transform. Inverse Probl. 27, 055006 (2011)
20. Graham, M., Gibbs, J., Cornish, D., Higgins, W.: Robust 3-D airway tree segmentation for image-guided peripheral bronchoscopy. IEEE T. Med. Imag. 29, 982–997 (2010)
21. Helou Neto, E., De Pierro, A.R.: Incremental subgradients for constrained convex optimization: A unified framework and new methods. SIAM J. Opt. 20, 1547–1572 (2009)
22. Helou Neto, E., De Pierro, A.R.: On perturbed steepest descent methods with inexact line search for bilevel convex optimization. Optimization 60, 991–1008 (2011)
23. Herman, G.T., Davidi, R.: On image reconstruction from a small number of projections. Inverse Probl. 24, 045011 (2008)
24. Herman, G.T., Garduño, E., Davidi, R., Censor, Y.: Superiorization: An optimization heuristic for medical physics. Med. Phys. 39, 5532–5546 (2012)
25. Holdsworth, C., Kim, M., Liao, J., Phillips, M.: A hierarchical evolutionary algorithm for multiobjective optimization in IMRT. Med. Phys. 37, 4986–4997 (2010)
26. Lauzier, P., Tang, J., Chen, G.: Prior image constrained compressed sensing: Implementation and performance evaluation. Med. Phys. 39, 466–480 (2012)
27. Levitan, E., Herman, G.T.: A maximum a posteriori probability expectation maximization algorithm for image reconstruction in emission tomography. IEEE T. Med. Imag. 6, 185–192 (1987)
28. Men, C., Romeijn, H., Jia, X., Jiang, S.: Ultrafast treatment plan optimization for volumetric modulated arc therapy (VMAT). Med. Phys. 37, 5787–5791 (2010)

29. Nesterov, Y.: Introductory Lectures on Convex Optimization. Kluwer (2004)
30. Nikazad, T., Davidi, R., Herman, G.T.: Accelerated perturbation-resilient block-iterative projection methods with application to image reconstruction. Inverse Probl. 28, 035005 (2012)
31. Nurminski, E.: Envelope stepsize control for iterative algorithms based on Fejer processes with attractants. Optimiz. Method. Softw. 25, 97–108 (2010)
32. Penfold, S., Schulte, R., Censor, Y., Rosenfeld, A.: Total variation superiorization schemes in proton computed tomography image reconstruction. Med. Phys. 37, 5887–5895 (2010)
33. Rardin, R., Uzsoy, R.: Experimental evaluation of heuristic optimization algorithms: A tutorial. J. Heuristics 7, 261–304 (2001)
34. Ruszczynski, A.: Nonlinear Optimization. Princeton University Press (2006)
35. Scheres, S., Gao, H., Valle, M., Herman, G.T., Eggermont, P., Frank, J., Carazo, J.M.: Disentangling conformational states of macromolecules in 3D-EM through likelihood optimization. Nat. Methods 4, 27–29 (2007)
36. Shepp, L., Vardi, Y.: Maximum likelihood reconstruction for emission tomography. IEEE T. Med. Imag. 1, 113–122 (1982)
37. Sidky, E., Pan, X.: Image reconstruction in circular cone-beam computed tomography by constrained, total-variation minimization. Phys. Med. Biol. 53, 4777–4807 (2008)
38. Studholme, C., Hill, D., Hawkes, D.: Automated three-dimensional registration of magnetic resonance and positron emission tomography brain images by multiresolution optimization of voxel similarity measures. Med. Phys. 24, 25–35 (1997)
39. Wernisch, L., Hery, S., Wodak, S.: Automatic protein design with all atom force-fields by exact and heuristic optimization. J. Mol. Biol. 301, 713–736 (2000)
40. Wu, Q., Mohan, R.: Algorithms and functionality of an intensity modulated radiotherapy optimization system. Med. Phys. 27, 701–711 (2000)
41. Yeo, B., Sabuncu, M., Vercauteren, T., Holt, D., Amunts, K., Ziles, K., Goland, P., Fischl, B.: Learning task-optimal registration cost functions for localizing cytoarchitecture and function in the cerebral cortex. IEEE T. Med. Imag. 29, 1424–1441 (2010)
42. Zanakis, S., Evans, J.: Heuristic optimization - why, when, and how to use it. Interfaces 11, 84–91 (1981)
43. Zhang, X., Wang, J., Xing, L.: Metal artifact reduction in x-ray computed tomography (CT) by constrained optimization. Med. Phys. 38, 701–711 (2011)

Parallel Algorithms for Combinatorial Pattern Matching

Valentin E. Brimkov

Mathematics Department, SUNY Buffalo State College, Buffalo, NY 14222, USA
brimkove@buffalostate.edu

Abstract. In this talk we discuss on parallel computation approach to two-dimensional combinatorial pattern matching. The latter features numerous applications in pattern recognition, low level image processing, computer vision and, more recently, multimedia. After introducing some basic notions and concepts and recalling related key facts, we briefly discuss the basic steps of a parallel algorithm design, illustrating them by author's results on the problem of detecting all two-dimensional repetitions in a two-dimensional array.

Keywords: Combinatorial pattern matching, two-dimensional array, tandem, parallel algorithm, CRCW-PRAM, divide and conquer.

1 Introduction

The classical pattern matching problem consists of finding all occurrences of an array called the *pattern* in another array called the *text*. Along with it, a variety of germane problems have been studied over the years. Particularly, important tasks have been seen in identifying interesting patterns, motifs, or any kind of meaningful features in a given text. Notable among these are repetitions of certain portions of a text. Studying repetitions, periodicities, and other similar regularities play a central role in various facets of computer science and subtend properties that are either interesting and deep from the standpoint of combinatorics, or susceptible to exploitation in tasks of automated inference and data compression, or both. Such studies date back at least to the beginning of the last century, and may be classified in part as computer science "ante litteram" [28,29]. In the last decades among the most important motivations of the discipline is its relevance to computational biology, and more precisely, to the automated analysis of biosequences.

More recently, efforts have been made at generalizing clever string searching techniques, notions and results to structures of higher dimensions, particularly two-dimensional arrays, where texts and patterns can be considered as "bit-map" images, represented by matrices of pixels and stored in a database. This is naturally driven by various applications to *pattern recognition*, low level *image processing*, *computer vision* and, more recently, *multimedia*. Like with strings, efficient solutions in higher dimensions rest heavily on defining, classifying and studying the possible periodicities in two-dimensional arrays (see [1,2,15,21,22]). Consequently,

R.P. Barneva, V.E. Brimkov, and J. Šlapal (Eds.): IWCIA 2014, LNCS 8466, pp. 8–16, 2014.

two-dimensional periodicities and related notions have been brought about and studied in this context.

Combinatorial pattern matching problems have been studied in the framework of diverse computation models and settings. Numerous algorithms exist (mostly for problems on strings), including optimal sequential algorithms, online algorithms, algorithms that solve a problem in real time, or ones that use only constant auxiliary space in addition to the input. In this talk we focus on parallel computation approach to pattern matching problems. That approach appears to be quite suitable regarding the very nature of the considered problems. Moreover, the author believes that the outlined techniques can be applied to other problems of combinatorial image analysis.

After introducing some basic notions and concepts and recalling related key facts, we briefly discuss the basic steps of a parallel algorithm design, illustrating them by author's results on the problem of detecting all two-dimensional repetitions in a two-dimensional array.

2 CRCW-PRAM and Other Models of Parallel Computation

Below we recall some basic points related to parallel models of computation. For a detailed accounting of the matter the reader is referred to [19,20].

PRAM is a synchronous model of parallel computation. It is a natural generalization of the random access memory model (RAM), which is the model commonly used in sequential computation. PRAM consists of m shared memory cells $M_1, \ldots, M_m$ and p processors $P_1, \ldots, P_p$. Each processor can be considered as a RAM with a private local memory. Processors communicate via the shared memory. Thus, restricting the amount of the shared memory corresponds to restricting the amount of information which can be communicated between processors in one step. A step of computation consists of read, compute and write phases. In the *read phase* every processor may read from a shared memory cell. In the *compute phase* every processor may perform a local operation. In the *write phase* every processor may write to one shared memory cell. Any number of processors can simultaneously read from or attempt to simultaneously write to the same shared memory cell.

Assume now that an input x consists of n values $x_1, \ldots, x_n$, where $n \leq m$. Initially, these values are located in shared memory cells $M_1, \ldots, M_n$. If at the end of the computation there are $n' \leq m$ output values, they appear in the first n' shared memory cells.

Work of a PRAM-algorithm is defined as the product of its running time and the number of processors used. A parallel algorithm is said to be *optimal* if its work equals the running time of the fastest possible sequential algorithm for the considered problem.

The following classical result is often used in the evaluation of the time and work performance of parallel algorithms.

Lemma 1. *(Brent's Lemma [10]) Any parallel algorithm of time t that consists of a total of x elementary operations can be implemented on p processors in $\lceil \frac{x}{p} \rceil + t$ time.*

Lemma 1 implies that the work of any PRAM-algorithm solving a problem is at least constant factor times the sequential complexity of the problem.

An important issue in CRCW-PRAM is how to resolve the write conflict, when a number of processors are concurrently attempting to write to the same shared memory cell. This leads to several variants of CRCW-PRAM (see [19] for a survey). Some of the most important are the PRIORITY, ARBITRARY and COMMON models. In the COMMON model [24], more than one processor is allowed to simultaneously write to the same memory cell only if they are all writing a common value. In the ARBITRARY model [31], if more than one processor attempts to write to the same memory cell, an arbitrary one succeeds. In the PRIORITY model [23] processors are assigned distinct priorities, and if more than one processor attempts to write to the same memory cell, the one with the highest priority succeeds.

It is well-known (and also easy to see) that any algorithm that works within the PRIORITY model will work without changes within the ARBITRARY model. Hence, the PRIORITY model is at least as powerful as the ARBITRARY model. Similarly, one can see that the ARBITRARY model is at least as powerful as the COMMON model. We remark that the algorithms discussed in the present paper work in the framework of the COMMON model which is the weakest version of CRCW-PRAM. Thus, they work without any change also within the PRIORITY and ARBITRARY models. Furthermore, the complexity bounds which hold for the COMMON model apply to the ARBITRARY and PRIORITY models as well.

In designing parallel algorithms for searching an array often is used the well-known integer minima algorithm of Fich, Ragde and Wigderson [20], denoted further the FRW-algorithm. It can be applied for various purposes, whenever it is necessary to find in a string of size n, in constant CRCW-PRAM time with n processors, the minimal/maximal position containing certain value. For example, the FRW-algorithm can be used to find in constant time with n processors the minimal integer in a set of n integers in the range $1, \ldots, n$, or to determine the first occurrence of a string in another string (see [4]).

3   Using the Divide and Conquer Strategy: An Example

The characterization and detection of repetitive structures arises in a variety of applications. In strings, the notions of a *repetition* or *square*, i.e., two consecutive occurrences of a same primitive[1] word, is intermingled with the dawn of language theory and combinatorics on strings. Long after A. Thue proved the existence of indefinitely long "square-free" strings over an alphabet of 3 symbols or more

---

[1] A word S is *primitive* if $S = uu \ldots u = u^k$ where k is some positive integer, implies $k = 1$ and $S = u$.

[28,29], optimal $O(n \log n)$ work serial and parallel algorithms were developed to detect all squares in a string as well as for some related applications (see, e.g., [3,4,7,8,16,17,25]).

Comparatively less has been done about two-dimensional extensions of the notions of square and repetition. In [5], a two-dimensional repetition called a *tandem* is defined as a configuration consisting of two occurrences of the same (primitive) block that touch each other with one side or corner. Being primitive for a block means that it cannot be expressed itself by repetitive placement of another block.

In what follows we sketch the basic steps of *optimal* algorithms to *detect* all positioned tandems in an array. The algorithm uses a divide and conquer strategy as well as some basic results of combinatorial pattern matching.

3.1 Basic Definitions and Facts

Let Σ be a finite *alphabet* of *symbols*. A *two-dimensional array* (or *2D array*, for short) on Σ is any $m \times n$ rectangular array $X[0..m-1, 0..n-1]$ with $m > 1$ *rows* and $n > 1$ *columns*. Any rectangular sub-array of X is a *block*. An array X is *primitive* if it cannot be partitioned into non-overlapping replicas of some block W, i.e., if setting $X = \begin{array}{|ccc|} \hline W & \cdots & W \\ \cdots & \cdots & \cdots \\ W & \cdots & W \\ \hline \end{array}$, where X has k rows and l columns, implies $k = 1$, $l = 1$. Furthermore, X is *vertically* (resp. *horizontally*) *primitive* if it cannot be represented in the form $X = \begin{array}{|c|} \hline W \\ \cdots \\ W \\ \hline \end{array}$ (resp. $X = \boxed{W \cdots W}$), for some block $W \neq X$.

Given a 2D array X, a *tandem* in X is a configuration consisting of two occurrences of the same primitive block W that touch each other along one side (tandem Type 1) or at a corner (tandem Type 2).

Block W is called the *root* of the tandem. If W is vertically (resp., horizontally) primitive, X is called a *vertical* (respectively, *horizontal*) *tandem*. Clearly, X is a tandem if and only if it happens to be simultaneously a vertical and a horizontal tandem.

Tight bounds for the number of tandems in an array are given by the following theorem.

Theorem 1 [5] *An $n \times n$ array can contain $\Theta(n^3 \log_\Phi n)$ tandems of Type 1 and $\Theta(n^4)$ tandems of Type 2, where $\Phi = \frac{1+\sqrt{5}}{2}$ is the golden ratio.*

More details about tandems are found in [5].

3.2 Algorithm's Outline

We have the following theorem.

Theorem 2 *There is an optimal, optimally fast $O(\log \log n)$ time CRCR-PRAM algorithm for detecting all tandems in an $n \times n$ array A. The algorithm finds all tandems of Type 1 with $O(n^3 \log n)$ work and all tandems of Type 2 with $O(n^4)$ work.*

In [12] the above theorem is proved by exhibiting a tandem detection algorithm with the stated time and work complexity. The discussion below refers to detection of tandems of Type 1 (see [6] for a serialized version).

The algorithm uses as a subroutine the optimal, optimally fast parallel two-dimensional pattern matching algorithm of Cole et al. [15], denoted further by OptPM. It preprocesses an $m \times m$ pattern in *optimal* $O(\log \log m)$ time with $\frac{m^2}{\log \log m}$ processors, and after that it searches for occurrences of the pattern in the $n \times n$ text in *constant time* with n^2 processors. The algorithm also frequently uses the integer minima FRW-algorithm.

While our tandem detection algorithm results in part from a prudent orchestration of serialized routines of the square detection algorithm from [4], it differs substantially from that one, in that most of the work here is to cope with the somewhat subtle interplay between horizontal and vertical repetitive structures. For one thing, that algorithm does not need to preprocess the whole text, while in our approach the entire array is preprocessed, and this is followed by $\lceil \log^2 n \rceil$ additional stages.

The algorithm consists of two phases: preprocessing phase and search (stage) phase.

1. Preprocessing
In this phase, the algorithm calls as a subroutine the preprocessing phase of the Cole et al. Opt. PM. Through it, the array A is preprocessed and, as a result, an array Wit of *witnesses*[2] for A is computed. Informally speaking, that is a table containing information (later used in the search phase) about the positions where mismatches occur between two superimposed copies of an array.

2. Search Phase (Stages)
The stages are indexed by pairs (η_1, η_2), where $0 \le \eta_1 \le \lceil \log n \rceil - 1$, $0 \le \eta_2 \le \lceil \log n \rceil - 1$. For definiteness, we assume $\eta_1 \ge \eta_2$. We look for tandems of the form $T = \boxed{W|W}$, the case where $T = \boxed{\dfrac{W}{W}}$ being symmetric. The operation of the stage is as follows. First, we detect all repetitions of the form $X = \boxed{W|W}$ where X is an array of size compatible with the stage index; after that, we test which of these are tandems. For this, we need to solve a set of 2D pattern matching problems.

More into details, consider stage number (η_1, η_2) and assume for generality that $\eta_1, \eta_2 \ne 0$. The algorithm looks for repetitions of the form $\boxed{W|W}$ where W is an $w_1 \times w_2$ array, such that $2l_{\eta_1} - 1 \le w_1 < 2l_{\eta_1+1} - 1$, $2l_{\eta_2} - 1 \le w_2 < 2l_{\eta_2+1} - 1$, $l_{\eta_1} = 2^{\eta_1}$, $l_{\eta_2} = 2^{\eta_2}$. Applying "divide and conquer" strat-

---

[2] To get acquainted with the powerful concepts of witnesses and duels the reader is referred to [30] and [1].

egy, we partition A into $\frac{n}{l_{\eta_1}} \times \frac{n}{l_{\eta_2}} = \frac{n^2}{l_{\eta_1} l_{\eta_2}}$ blocks of size $l_{\eta_1} \times l_{\eta_2}$. Let $B = A\,[P_1..P_1 + l_{\eta_1} - 1, P_2..P_2 + l_{\eta_2} - 1]$ be a block of the partition and T a tandem of the required size, whose left root fully contains B. Then we say that T is *hinged* on B.

For every block B, we next perform a *substage*, with every substage consisting of two *rounds*, respectively called Round 1 or "horizontal round", and Round 2 or "vertical round". The task of Round 1 is to detect all horizontal tandems, while that of Round 2 is to certify those among them which are also vertical tandems.

Both rounds are quite technical, including a number of lemmas and numerous constructions and procedures. The main difficulty is how to perform the substages with a sufficiently small amount of operations and processors, so that the entire algorithm to be optimally fast, and also when summed up, the total amount of work to feature an optimal parallel algorithm. In this regard, crucial appeared to be the use (in many different occasions) of the integer minima FRW-algorithm and the search phase of the Cole et al. pattern matching algorithm, whose execution takes constant time. As a result, it has been shown that each substage of stage number (η_1, η_2) takes $O(l_{\eta_1}^2 l_{\eta_2})$ work. Detailed description of the algorithm is available in [12].

3.3 Algorithm's Time and Work

The preprocessing phase of the algorithm takes $O(\log\log n)$ time and $O(n^2)$ work. All stages are executed simultaneously, as every stage takes constant time, i.e., the entire stage phase is executed in constant time.

To estimate the overall amount of algorithm's work, consider firs stage number (η_1, η_2). There are $O(\frac{n^2}{l_{\eta_1} l_{\eta_2}})$ substages in that stage.

As mentioned above, one substage takes $O(l_{\eta_1}^2 l_{\eta_2})$ work. Then the total work of the stage phase amounts to

$$O\left(\sum_{\eta_1=1}^{\log n} \sum_{\eta_2=1}^{\log n} \frac{n^2}{l_{\eta_1} l_{\eta_2}} \cdot l_{\eta_1}^2 l_{\eta_2}\right) = O\left(\sum_{\eta_1=1}^{\log n} \sum_{\eta_2=1}^{\log n} n^2 l_{\eta_1}\right) =$$

$$= O\left(\sum_{\eta_1=1}^{\log n} n^2 \log n . l_{\eta_1}\right) = O\left(n^3 \log n\right).$$

Using Brent's Lemma, one can slow down the stage phase to $O(\log\log n)$ time and $O(n^2 \log n)$ work. Then all independent stages can be executed simultaneously in $O(\log\log n)$ time and $O(n^3 \log n)$ work with $\frac{n^3 \log n}{\log\log n}$ processors. Thus the overall time and work of the algorithm amount to $O(\log\log n)$ and $O(n^3 \log n)$, respectively.

Since the overall work of the algorithm matches the lower bound for the number of horizontal tandems of Type 1 (see Theorem 1), it follows that the algorithm is optimal.

Moreover, the algorithm has the same time complexity as the one from [4], which is proved to be optimally fast parallel CRCR-PRAM square-freedom testing algorithm. Hence, the same applies to the tandem detection algorithm of Theorem 2.

4   Concluding Remarks

In this note we have discussed on parallel algorithms for certain pattern matching problems in two dimensions. We sketched an optimal, optimally fast parallel algorithm for detecting all positioned tandems in an $(n \times n)$-array. The computation takes optimal $O(\log \log n)$ CRCR-PRAM time and $O(n^3 \log n)$ work for tandems of Type 1 and $O(n^4)$ work for tandems of Type 2. Note that it is possible to test the tandem-freedom of an $(n \times n)$-array in optimal $O(\log \log n)$ time and $n^2 \log n$ work for tandems of Type 1 and with $O(n^2 \log^2 n)$ work for tandems of Type 2 [11]. Thus, in 2D, the exhaustive discovery appears to be a harder problem than the simple test for freedom, in contrast with what happens in the linear case of strings, where both problems require in the general case the same optimal $O(\log \log n)$ time and $O(n \log n)$ work.

We believe that the presented algorithm can be utilizing for the purposes of compressed encoding of 2D arrays, in particular those representing digital planes.

Studies in combinatorial pattern matching and the related theory of words interfere with research on digital lines and planes in image analysis and digital geometry. For instance, the well-known Sturmian words represent discretizations of irrational straight lines (see, e.g., [26]). In recent years, 2D extensions of Sturmian words and related digital flatness matters have been studied by several authors (see, e.g., [9,13,14,18,32] and the bibliography therein). Any digital plane can be represented by an infinite 2D array on two- or three-letter alphabets. The periodicity types of binary plane representations have been recently studied in [13] by employing well-known definitions of 2D periodicity and related results from [1].

An important problem in image analysis is to recognize whether a given set of points constitutes a portion of a digital plane. In [27], an algorithm is presented which solves the problem by exploiting certain properties of "combinatorial pieces" of digital planes that appear to be one-dimensional Sturmian words. As discussed in [9,32], some properties of 2D Sturmian words might be instrumental in designing truly two-dimensional algorithms for digital plane recognition. In this regard, a challenging task is to look for properties that characterize 2D arrays corresponding to digital planes.

References

1. Amir, A., Benson, G.: Two-dimensional periodicity and its applications. In: Proc. 3rd ACM-SIAM Symp. on Discrete Algorithms, pp. 440–452 (1992)
2. Amir, A., Benson, G., Farach, M.: An alphabet independent approach to two-dimensional pattern matching. SIAM J. Comput. 23(2), 313–323 (1994)

3. Apostolico, A.: Optimal parallel detection of squares in strings. Algorithmica 8, 285–319 (1992)
4. Apostolico, A., Breslauer, D.: An optimal $O(\log \log n)$ time parallel algorithm for detecting all squares in strings. SIAM J. Comput. 25(6), 1318–1331 (1996)
5. Apostolico, A., Brimkov, V.E.: Fibonacci arrays and their two-dimensional repetitions. Theoret. Comput. Sci. 237, 263–273 (2000)
6. Apostolico, A., Brimkov, V.E.: Optimal discovery of repetitions in 2D. Discrete Applied Mathematics 151, 5–20 (2005)
7. Apostolico, A., Preparata, F.P.: Optimal off-line detection of repetitions in a string. Theoret. Comput. Sci. 22, 297–315 (1983)
8. Apostolico, A., Preparata, F.P.: Structural properties of the string statistic problem. J. Comput. Systems 31(3), 394–411 (1985)
9. Berthé, V., Vuillon, L.: Tilings and rotations on the torus: a two-dimensional generalization of Sturmian words. Discrete Mathematics 223, 27–53 (2000)
10. Brent, R.P.: Evaluation of general arithmetic expressions. J. ACM 21, 201–206 (1974)
11. Brimkov, V.E.: Optimally fast CRCW-PRAM testing 2D-arrays for existence of repetitive patterns. International Journal of Pattern Recognition and Artificial Intelligence 15(7), 1167–1182 (2001)
12. Brimkov, V.E.: Optimal parallel searching an array for certain repetitions. E.N. Discr. Math. 12 (2003), http://www.elsevier.nl/gej-ng/31/29/24/show/Products/notes/contents.htt
13. Brimkov, V.E., Barneva, R.P.: Plane digitization and related combinatorial problems. Discrete Applied Mathematics 147, 169–186 (2005); See also: Brimkov, V.E.: Digital flatness and related combinatorial problems. CITR-TR-120, University of Auckland, New Zealand, 45 pages (2002), http://www.citr.auckland.ac.nz/techreports/?year=2002
14. Cassaigne, J.: Two-dimensional sequences of complexity $mn + 1$. J. Automatic Language Combinatorics 4, 153–170 (1999)
15. Cole, R., Crochemore, M., Galil, Z., Gąsieniec, L., Hariharan, R., Muthukrishnan, S., Park, K., Rytter, W.: Optimally fast parallel algorithms for preprocessing and pattern matching in one and two dimensions. In: Proc. 34th IEEE Symp. Found. Computer Science, pp. 248–258 (1993)
16. Crochemore, M.: An optimal algorithm for computing the repetitions in a word. Inform. Process. Let. 12(5), 244–250 (1981)
17. Crochemore, M.: Transducers and repetitions. Theoret. Comput. Sci. 12, 63–86 (1986)
18. Epifanio, C., Koskas, M., Mignosi, F.: On a conjecture on bidimensional words. Theoretical Computer Science 299(1-3), 123–150 (2003)
19. Fich, F.: The complexity of computation on the parallel random access machine (Chapter 21). In: Reif, J. (ed.) Synthesis of Parallel Algorithms. Morgan Kaufmann (1993)
20. Fich, F., Ragde, P., Wigderson, A.: Relations between CR-models of parallel computations. SIAM J. Comput. 17(3), 606–627 (1988)
21. Galil, Z., Park, K.: Truly alphabet-independent two-dimensional pattern matching. In: Proc. 33rd IEEE Symp. Found. Computer Science, pp. 247–256 (1992)
22. Galil, Z., Park, K.: Alphabet-independent two-dimensional witness computation. Manuscript (1993)
23. Goldschlager, L.: A unified approach to models of synchronous parallel machines. J. ACM 29, 1073–1086 (1982)

24. Kucera, L.: Parallel computation and conflicts in memory access. Inform. Process. Lett. 14(2), 93–96 (1982)
25. Main, G.M., Lorentz, R.J.: An $O(n \log n)$ algorithm for finding all repetitions in a string. J. Algorithms 5, 422–432 (1984)
26. Morse, M., Hedlund, G.A.: Symbolic dynamics II: Sturmian sequences. Amer. J. Math. 61, 1–42 (1940)
27. Réveillès, J.-P.: Combinatorial pieces in digital lines and planes. In: Proc. of the SPIE Conference "Vision Geometry IV", San Diego, CA, vol. 2573, pp. 23–34 (1995)
28. Thue, A.: Über unendliche Zeichenreihen. Norske Vid. Selsk. I Mat. Natur. Kl. Skr., Christiania (7), 1–22 (1906)
29. Thue, A.: Über die gegenseitige Lage gleicher Zeichenreihen. Norske Vid. Selsk. I Mat. Natur. Kl. Skr., Christiania (1), 1–67 (1912)
30. Vishkin, U.: Optimal parallel pattern matching in strings. Information and Control 67, 91–113 (1985)
31. Vishkin, U.: Implementation of simultaneous memory access in models that forbid it. J. Algorithms 4, 45–50 (1983)
32. Vuillon, L.: Local configurations in a discrete plane. Bull. Belgium Math. Soc. 6, 625–636 (1999)

Role of Image Processing
in Solar Coronal Research

Shadia Rifai Habbal[1], Miloslav Druckmüller[2], and Huw Morgan[3]

[1] Institute for Astronomy, University of Hawaii
shadia@ifa.hawaii.edu
[2] Institute of Mathematics, Faculty of Mechanical Engineering,
Brno University of Technology
druckmuller@fme.vutbr.cz
[3] Institute of Mathematics,
Physics and Computer Science, Aberystwyth University
hum2@aber.ac.uk

Abstract. The wealth of information regarding the structure and distribution of magnetic fields and plasmas embedded in images of the solar corona taken in the visible wavelength range and/or the extreme ultraviolet (EUV), covers several orders of magnitude in brightness, in the radial and azimuthal directions. As such, they present serious visualization challenges. These can only be revealed with the use of image processing tools. This presentation will focus on results from two recently developed approaches: (1) The Adaptive Circular High-pass Filter, ACHF, and the Normalized Radial Gradient Filter, NRGF, which are ideal for limb observations of the corona made in the visible wavelength range during total solar eclipses or with coronagraphs. (2) The Noise Adaptive Fuzzy Equalization (NAFE) and the multi-scale Gaussian normalization process (MGN), suitable for the visualization of fine structures in EUV images. These methods yield artifact-free images and uncover details that are hidden in the original unprocessed images. Such details have led to the discovery of new features that are essential for exploring the dynamics and thermodynamics of structures in the solar corona.

Keywords: Sun: corona, Ultraviolet emission, Visible emission, Image processing.

1 Introduction

The solar corona is the extension of the solar atmosphere outwards into interplanetary space, starting from a few megameters above the visible solar surface. With a characteristic temperature of at least 10^6 K, the coronal plasma is fully ionized, and expands with the solar magnetic field. This expansion is produced by the solar wind that shapes planetary environments. It is precisely the expansion of the coronal plasma that leads to a very sharp radial gradient in the intensity of the emission from particles excited by scattering, collisions and resonant excitation. In addition, a significant fraction of the plasma forming the

R.P. Barneva, V.E. Brimkov, and J. Šlapal (Eds.): IWCIA 2014, LNCS 8466, pp. 17–24, 2014.
© Springer International Publishing Switzerland 2014

solar atmosphere remains bound to the Sun by magnetic field lines assuming the shape of loops connecting regions with opposite magnetic polarities. These loops produce emission with a very large dynamic range as well. Hence, to reveal all the fine details of coronal structures, as seen by the naked eye during a total solar eclipse, one has to develop tools to handle the several orders of magnitude range of the steep radial and azimuthal gradients of the coronal emission.

At present, the two most common wavelength ranges for imaging the corona are the visible and the EUV. Observations in the visible naturally exclude the solar disk, which is dominated by photospheric emission and is a million times brighter than the corona in that wavelength range. The only opportunities available for these observations are during natural total solar eclipses, or with manmade coronagraphs. The EUV emission is slightly different, as it reveals structures both on the solar disk and off the limb. However, the impact of the radial gradient is even more striking in the emission observed off the limb, since this emission is proportional to the product of the electron and ion densities, and hence drops much more steeply with radial distance than the white light.

The role of any image processing technique applied to the corona is to reduce the radial gradient and enhance high spatial frequencies to increase the visibility of structures. The example shown in Figure 1 is a section of the corona imaged in white light during the total solar eclipse of 2008 August 1. With the exception of the prominences that appear as bright concentrations in a few places around the limb of the Sun, nothing else stands out. The inset shown at the top illustrates the details of coronal structures embedded in this image that become visible with the application of an image processing tool, the ACHF in this case, which will be described in more detail in the following section.

Fig. 1. Impact of image processing for bringing out details of coronal structures. The background image is from the total solar eclipse of 2006 March 29. The inset shows a section of the processed image by the ACHF.

The goal of this presentation is to illustrate the role of a few recently developed image processing techniques in revealing the bewildering details embedded in white light and EUV images of the corona. We give a brief description of the basic properties of these approaches, and present a few examples resulting from their application to coronal images. We conclude by discussing the role these processed images have played in reshaping our understanding of coronal structures, and their implications for the way we look at the solar corona, in both the visible and EUV wavelength range.

2 Tools and Examples of Image Processing Tools as Applied to Coronal Images

2.1 Tools Applicable to White Light and Visible Wavelength Images

The ACHF Method. The Adaptive Circular High-pass Filter, or ACHF, was developed by [2] to process total solar eclipse images. The goal is to visualize the fine details of the structures embedded in photographic images (whether film or digital), but naturally visible to the naked eye. Typically during a total solar eclipse, a sequence of different exposure times must be used to capture the very large dynamic range that the human vision is able to manage because of a different image acquisition principle. Unlike classical or digital photography the human vision is a differential analyzer which is not able to "measure" the absolute brightness but it is only able to compare the "pixel brightness" with its neighborhood. This principle has the great advantage that even extreme brightness ratios in the whole image do not prevent the human vision from seeing small low contrast details in the solar corona. To imitate this ability of the human vision is not simple. The first step is precise image alignment taken with different exposure times and the creation of one high dynamic range (HDR) image. The alignment method must be able to handle not only the shift between images, but rotation and scaling as well, because several different instruments must often be used for the creation of a single image. A modified phase-correlation method was developed for that purpose (see details in [2]).

The next step in the image processing is to expose the more than six orders of magnitude range of brightness in the resulting image. This is achieved with a discrete convolution of the image with a variable kernel, the properties of which are set according to the local properties of the image. The actual implementation of the approach is rather complex, as a number of parameters have to be applied intuitively. An example of the outcome of the application of this technique is shown in the inset of Figure 1.

The NRGF Method. The Normalizing Radial Gradient Filter [10], or NRGF, is a technique used primarily to remove the radial gradient in coronal images. The normalization of the emission is undertaken in concentric bands covering the full 360° position angles at different radial distances. Essentially, at every point in the image, a mean $I(r)_{<\phi>}$ is calculated over all position angles at distance r. It is then subtracted from the original intensity, $I(r, \phi)$, at distance r and position angle ϕ. The difference is then divided by the standard deviation $\sigma(r)_{<\phi>}$ of intensities calculated over all ϕ's at distance r. Hence the intensity at every point (r, ϕ) in a processed image is given by:

$$I'(r, \phi) = [I(r, \phi) - I(r)_{<\phi>}]/\sigma(r)_{<\phi>} \tag{1}$$

An example of the application of this technique to LASCO/C2 images is given in Figure 2.

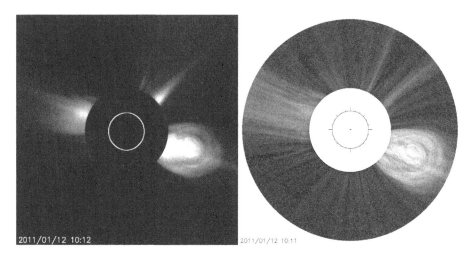

Fig. 2. LASCO/C2 observation of the corona during 2011/01/12 10:12. A large CME is seen in the south-west. Left - standard image processed by subtraction of a monthly-minimum background and histogram equalization. This image is typical of the standard used by the community. Right - the same observation processed by point filters, NRGF, and MGN. Note the great increase in detail in all regions of the corona. Plumes are clearly seen in dark coronal holes, and the CME's structural detail is greatly enhanced. This detail is crucial in analyzing the nature of CMEs, and their interaction with the background corona. Note that the streamers in the standard processing (left) appear to narrow with increasing distance from the Sun - this is an artifact of the processing, and is a grave misrepresentation of the true coronal structure, as correctly seen in the right image.

One of the limitations of this technique is its inability to compensate for structures with different contrast at the same radial distance. The Fourier Normalizing Radial Gradient Filter [5], or FNRGF, compensates for this limitation.

2.2 Tools Applicable to EUV Images

While the steep radial gradient is the main challenge for visualizing coronal emission off the limb of the Sun, with the EUV images it's the additional

challenge of the wide range of intensity levels across the solar disk. Two recent techniques have been developed to handle these images, primarily those
from the Atmospheric Imaging Assembly (AIA) [8] aboard the Solar Dynamic
Observatory (SDO) [11] which has the highest spatial resolution available to
date, of approximately 1". In those images, the steep contrast is produced between different loop-like magnetized density structures, making the visualization
of all structures, simultaneously, practically impossible.

The NAFE Method. For the AIA/SDO images, the high contrast features are
typically in the high spatial frequencies, which implies that the well-established
method of attenuating low spatial frequencies with techniques based on Fourier
transforms or convolutions are not applicable. On the other hand, the adaptive
histogram equalization approach [12], while potentially applicable, introduces
artifacts and amplifies noise. The Noise Adaptive Fuzzy Equalization method [3],
NAFE, essentially combines the two approaches to overcome the shortcomings of
each and produces artifact free images with unprecedented details. An example
of the application of the NAFE method to an AIA/SDO 171 Å image is shown
in Figure 3, with the unprocessed image on the left, and the processed one on
the right.

Fig. 3. Comparison of unprocessed AIA/SDO 171 Å image, $\gamma = 2.4$ (left) and NAFE
processed image $\gamma = 2.4$, $w = 0.2$, $\sigma = 12$ (right)

The MGN Method. The Multiscale Gaussian Normalization method, MGN,
[9] is similar to an adaptive histogram equalization. The basic approach is to first
take the square root of the image, B, to reduce the dominance of the contrast
range. A normalized image, C, is then computed by taking a 2 dimensional
Gaussian kernel, k_w, of width w pixels, subtracting the convolution of B with
the kernel, and dividing by an overall standard deviation (see details in [9]).
The advantage of this method over the NAFE method is that it is simpler to
implement. As a method using Gaussian smoothing, it can be easily programmed
with a few lines of code. This approach also treats the details of the corona that
are observed off the limb as well as on the disk.

The example of Figure 2 shows the result from the application of the NRGF and the MGN. The example of Figure 4 is a combination of the MGN method applied to the AIA/SDO 171 Å image superimposed on the ACFH-processed total solar eclipse white light image, both taken at the same time on 2010 July 11. Figure 4 is remarkable for two reasons. (1) It shows a perfect match between observations taken with two totally different instruments at the same time. (2) It also demonstrates the power and reliability of two different image processing techniques, the ACHF and MGN, whereby the finest details in both data sets match perfectly.

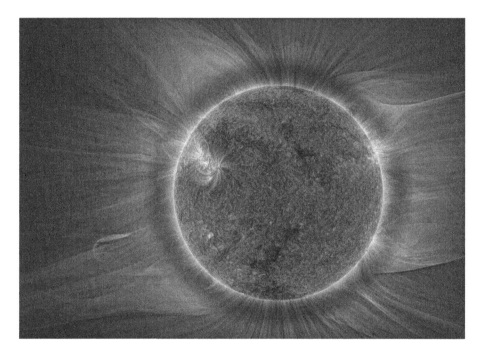

Fig. 4. Composite of an AIA/SDO image (solar disk) processed with MGN and white light eclipse image processed with the ACHF from observations on 2010 July 11. Note the perfect correspondence between the details of the two different instruments just above the solar limb.

3   Scientific Impact of Image Processing Techniques for the Physics of the Solar Corona

The processing of coronal images is not merely important for esthetic reasons. The information embedded in these images is fundamental for understanding the physical processes that control and shape the coronal plasmas. Application of the techniques described above have led to major discoveries. Examples of such discoveries are: the presence of localized ion density enhancements, with

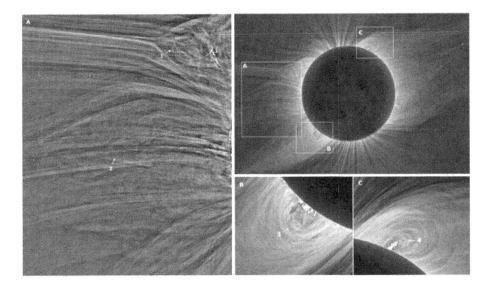

Fig. 5. Appearance of faint structures in the corona. The white light eclipse image (top right) was processed with the ACHF. The insets labeled A, B and C are shown in detail below. Features labeled by the arrows 1, 2, 3 and 4 refer to faint coronal structures in the form of expanding bubbles (see [4]).

implications for the heating of the heavier elements there [7]; the ubiquitous presence of nested loops surrounding filaments and forming the base of streamers [4,6]; and the presence of a new class of faint structures akin to vortices and instabilities [4], as shown in the example of Figure 5, to name a few.

4   Summary

Despite the abundance of image processing techniques, this presentation has focused on a few recently developed image processing tools that have led to breakthroughs in our understanding of some of the physical properties of the solar corona. It is the hope of the authors that these methods will achieve wide use with their application to the broad range of ground- and space-based coronal observations, not limited to total solar eclipses.

Acknowledgments. SRH appreciates the invitation to give this lecture, and acknowledges financial support from Brno University to attend this conference.

References

1. Brueckner, G.E., et al.: The Large Angle Spectroscopic Coronagraph. Solar Phys. 162, 357–402 (1995)
2. Druckmüller, M., Rušin, V., Minarovjech, M.: A New Numerical Method of Total Solar Eclipse Photography Processing. Contributions of the Astronomical Observatory Skalnate Pleso 36, 131–148 (2006)
3. Druckmüller, M.: A Noise Adaptive Fuzzy Equalization Method for Processing Solar Extreme Ultraviolet Images. Astrophys. J. Suppl. 2017, 25 (5pp) (2013)
4. Druckmüller, M., Habbal, S.R., Morgan, H.: Discovery of a New Class of Coronal Structures in White Light Eclipse Images. Astrophys. J. 785, 14 (7pp) (2014)
5. Druckmüllerová, H., Morgan, H., Habbal, S.R.: Enhancing Coronal Structures with the Fourier Normalizing-Radial-Graded Filter. Astrophys. J. 737, 88 (10pp) (2011)
6. Habbal, S.R., Druckmüller, M., Morgan, H., et al.: Total Solar Eclipse Observations of Hot Prominence Shrouds. Astrophys. J. 719, 1362–1369 (2010)
7. Habbal, S.R., et al.: Localized Enhancements of Fe^{+10} Density in the Corona as Observed in Fe XI 789.2 nm during the 2006 March 29 Total Solar Eclipse. Astrophys. J. 663, 598–609 (2007)
8. Lemen, J.R., et al.: The Atmospheric Imaging Assembly (AIA) on the Solar Dynamics Observatory (SDO). Solar Phys. 172, 17–40 (2012)
9. Morgan, H., Druckmüller, M.: Multi-scale Gaussian Normalization for Image Process: Application to Solar EUV Observations. Solar Phys. (in press, 2014)
10. Morgan, H., Habbal, S.R., Woo, R.: The Depiction of Coronal Structure in White Light Images. Solar Phys. 236, 263–272 (2006)
11. Pesnell, W.D., Thompson, B.J., Chamberlain, P.C.: The Solar Dynamics Observatory. Solar Phys. 275, 3–15 (2012)
12. Pratt, W.K.: Digital Image Processing, 4th edn. John Wiley & Sons (2004)

On Intersection Graphs of Convex Polygons

Valentin E. Brimkov[1], Sean Kafer[2],
Matthew Szczepankiewicz[2], and Joshua Terhaar[1]

[1] Mathematics Department, SUNY Buffalo State College, Buffalo, NY 14222, USA
brimkove@buffalostate.edu, terhaajb01@mail.buffalostate.edu
[2] Mathematics Department, University at Buffalo, Buffalo, NY 14260-2900, USA
{seankafe,mjszczep}@buffalo.edu

Abstract. Since an image can easily be modeled by its adjacency graph, graph theory and algorithms on graphs are widely used in image processing. Of particular interest are the problems of estimating the number of the maximal cliques in a graph and designing algorithms for their computation, since these are found relevant to various applications in image processing and computer graphics. In the present paper we study the maximal clique problem on intersection graphs of convex polygons, which are also applicable to imaging sciences. We present results which refine or improve some of the results recently proposed in [18]. Thus, it was shown therein that an intersection graph of n convex polygons whose sides are parallel to k different directions has no more than n^{2k} maximal cliques. Here we prove that the number of maximal cliques does not exceed n^k. Moreover, we show that this bound is tight for any fixed k. Algorithmic aspects are discussed as well.

Keywords: Intersection graph, Maximal clique, Convex polygon, Homothetic polygons.

1 Introduction

Graph theory and algorithms on graphs are widely used in image processing, as an image can easily be modeled by its adjacency graph [17]. The present paper is concerned with a class of intersection graphs. Given a family of sets S, its *intersection graph* has a vertex set labeled by the elements of S, and vertices $v_1, v_2 \in S$ are joined by an edge if and only if $v_1 \cap v_2 \neq \emptyset$ (see Figure 1). Intersection graph theory originated about half a century ago from works by Szpilrajn-Marczewski [25], Čulík [5], and Erdős, Goodman, and Pósa [6]. Since then, a substantial body of literature has been developed on the subject. Various structural results are available in the recent monograph [24], and algorithmic issues are discussed in [11]. Intersection graphs find numerous applications in different domains of scientific research, in particular in imaging sciences. A common application is in visualizing the relations among objects in a scene. This includes, for instance, drawing Euler diagrams [29], data visualization for the purposes of software engineering [9], video database queries [35], and visualization of overlapping clusters on graphs [30]. Intersection graphs have been used

R.P. Barneva, V.E. Brimkov, and J. Šlapal (Eds.): IWCIA 2014, LNCS 8466, pp. 25–36, 2014.
© Springer International Publishing Switzerland 2014

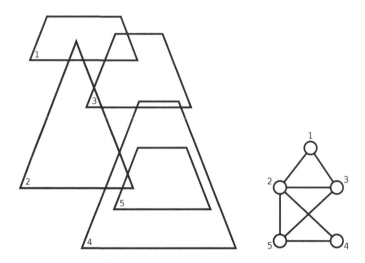

Fig. 1. Five polygons (left) and their intersection graph (right)

for graph-based recognition and visualization of grid patterns in street networks [12, 31], of contingency tables [32], and of intersections of solids [26]. Intersection graph-based discrete models of continuous n-dimensional spaces have been developed to address certain problems appearing in surface graphics [7].

It is well-known that every graph is an intersection graph of a certain family of sets [25]. However, various classes of intersection graphs possess properties that do not hold for general graphs. Thus, some classical NP-hard problems, like clique number, independence number, or chromatic number computation, admit polynomial algorithms on intersection graphs of specific families of sets. Efficient algorithms for such kinds of problems are applicable to problems of computer graphics and image analysis. For example, maximal clique computation has been applied to deformable object model learning in image analysis [36], image segmentation [14], and in other problems of image processing [28]. Applications of the maximal clique problem in computer vision are discussed in [2]. Maximal clique computation on certain intersection graphs is also relevant to some problems of computer graphics [27] and for visualization purposes in natural sciences [10].

Intersection graphs of convex polygons (either of special types, such as triangles and rectangles, or arbitrary) are among the most extensively studied and widely applicable intersection graphs (see, e.g., [1, 3, 4, 8, 13, 15, 18–23, 33, 34]). Thus, it was recently shown that an intersection graph of a set of isosceles right triangles (called semi-squares) can have $\Omega(n^3)$ maximal cliques which can be found in $O(n^3)$ time [19]. The study has been motivated by possible applications to important problems of computational biology. Another recent work [18] shows that the intersection graph of n convex polygons whose sides are parallel to k different directions has no more than n^{2k} maximal cliques. Based on some refinements of the above result, the authors also obtain upper bounds for intersection graphs of homothetic convex polygons and tighter bounds for some particular cases. For example, if the polygons are homothetic trapezoids, the number of maximal cliques is bounded above by n^4 rather than by n^8. In a recent paper [3] we reduced this last bound to n^3 which turns out to be optimal.

In the present paper we consider the maximal clique problem for intersection graphs of sets of arbitrary convex polygons. As a main result, we show that the exponent in the above-mentioned upper bound n^{2k} from [18] can be halved, i.e., a bound n^k holds. Moreover, we show that this last bound is tight for any fixed k. This also implies improvement on the upper bound for intersection graphs of homothetic polygons.

In the rest of this section we recall some basic notions and introduce notations to be used throughout the paper. In Section 3 we present our main results. In Section 4 we propose results related to intersection graphs of homothetic polygons. We conclude with some final remarks in Section 5.

2 Some Notions and Notations

A *clique* of an undirected simple graph G is a complete subgraph of G. A clique is *maximal* if it is not a proper subgraph of another clique, and *maximum* if no other clique of G has a greater number of vertices. The cardinality of a maximum clique is the graph's *clique number*.

For most of the following notions we follow [18].

Given a family $\mathcal{F}$ of subsets of the Euclidean plane $\mathbb{R}^2$, $IG(\mathcal{F})$ denotes the intersection graph of $\mathcal{F}$. The family $\mathcal{F}$ is called an *intersection representation* of the graph $G = IG(\mathcal{F})$.

SEG graphs are the intersection graphs of straight line segments, and k-DIR graphs are SEG graphs with representation in which all segments are parallel to at most k directions.

Our main result is about a class of intersection graphs known as $k_{DIR}-CONV$ graphs which are defined as follows. Let $\mathcal{L}$ be the set of all distinct lines in the plane $\mathbb{R}^2$ going through the origin. Denote by $\binom{\mathcal{L}}{k}$ all k-tuples of such lines. Let $L = \{l_1, \ldots, l_k\} \in \binom{\mathcal{L}}{k}$ be such a k-tuple. Let $\mathcal{P}(L)$ be the family of all convex polygons P such that every side of P is parallel to some line from L. Denote by $k_{DIR(L)} - CONV$ the class of intersection graphs of polygons from $\mathcal{P}(L)$. Then define the class $k_{DIR} - CONV = \bigcup_{L \in \binom{\mathcal{L}}{k}} k_{DIR(L)} - CONV$.

Thus, informally speaking, $k_{DIR} - CONV$ is the class of intersection graphs of convex polygons whose all sides are parallel to at most k directions.

Given a set $P \subset \mathbb{R}^2$, P_{hom} denotes the class of intersection graphs of sets homothetic to P.

3 Improved Upper Bounds on the Number of Maximal Cliques of $k_{DIR} - CONV$ Graphs

Let G be a graph on n vertices that belongs to $k_{DIR} - CONV$. As mentioned in the Introduction, [18] provides an upper bound of n^{2k} of the number of maximal cliques in G. In this section we present improvements on that bound.

3.1   General Results

Let $\mathcal{F}$ be a set of n convex polygons and let $G = IG(F)$. Let $L = \{l_1, \ldots, l_k\}$ be the set of lines corresponding to the different directions of sides of polygons from $\mathcal{P}$. For every line $l_i \in L$, let $\mathbf{c}_i$ be a vector perpendicular to it. The choice of the two directions perpendicular to l_i is arbitrary. For each polygon $P \in \mathcal{F}$ we define two sets of its points q_i and q'_i, where a sweeping line in the direction of $\mathbf{c}_i$ hits the points in q_i before all other points in P and hits the points of q'_i after all other points in P (see Figure 2a). Note that q_i and q'_i will always either be a vertex of P or a side of P perpendicular to $\mathbf{c}_i$.

For our purposes, we will treat any vertex of P that is a q_i or q'_i as an in-finitesimally small edge which is perpendicular to c_i (and thus parallel to l_i). Specifically, we will be choosing its direction to be such that, when needed, it is perpendicular to $\mathbf{c}_i$. As such, even in the cases where q_i or q'_i are single points, we can formally consider the sets q_i and q'_i of all polygons in the graph to be parallel.

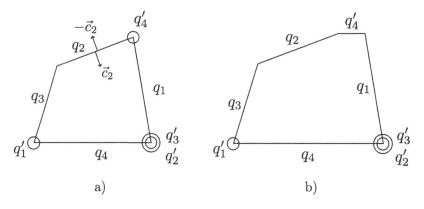

Fig. 2. Illustration of the sets q and q' for polygons where $k = 4$. Part a) includes a depiction of a $\mathbf{c}$ vector and its counterpart. Part b) provides an example where a q' set is a side.

Note that for some polygons in the graph representation a single vertex may be a member of many q or q' sets. Also note that since each side of each polygon either defines a new direction or is parallel to an existing one, all sides of any polygon are either q_i or q'_i for some i. See Figure 2 for illustration. Let $\mathbf{M}$ be a maximal clique in graph G, and for all i such that $l_i \in L$, let P_i be one of its polygons whose q_i is farther in the direction $\mathbf{c}_i$ than the q_i of any other polygon in $\mathbf{M}$. For every $P_i \in \mathbf{M}$, we will refer to its set q_i as b_i. Now, let $\mathbf{V}(b_1, \ldots, b_k)$ be the set of all polygons in the graph representation such that, for each i, its q_i is not strictly farther than b_i in the direction of $\mathbf{c}_i$ and its q'_i is not strictly farther than b_i in the direction of $-\mathbf{c}_i$. We have the following proposition.

Proposition 1. $M = V(b_1, \ldots, b_k)$

Proof: $M \subseteq V(b_1, \ldots, b_k)$: Let Q be in M. By definition, q_i of Q is not farther than b_i in the direction of c_i. Assume, for contradiction, that q_i' of Q is strictly farther than b_i in the direction of $-c_i$. If q_i' of Q contains the points of Q furthest in the direction of c_i, and b_i contains the furthest points of P_i in the direction of $-c_i$, yet b_i is still farther than q_i' of Q in the direction of c_i, then all points of Q must be strictly farther in the direction of $-c_i$ than all points of P_i, meaning that $Q \cap P_i = \emptyset$. This contradicts the fact that they are both in M.

$M \supseteq V(b_1, \ldots, b_k)$: The proof of this inclusion makes use of the following folklore fact (see [18]).

Fact 1 *Let P_1 and P_2 be convex polygons with $P_1 \cap P_2 = \emptyset$. Then P_1 and P_2 can be separated by a straight line containing a side of one of them.*

Let Q and R be in $V(b_1, \ldots, b_k)$. Assume, for contradiction, that $Q \cap R = \emptyset$. By Fact 1, Q and R are separated by a line containing one of their sides, say the side of R that is q_i of R. (As before, if q_i is a point, we consider it to be an infinitesimally small edge perpendicular to c_i.) Note that since the direction of c_i is arbitrary, provided it is perpendicular to line l_i, we can choose its direction to be such that this side of R is q_i and not q_i'. Since q_i of R is therefore strictly farther than all points of Q in the direction of c_i, it is not hard to see that b_i is strictly farther than all points of Q in the direction of c_i. Thus, as such q_i' of Q is strictly farther than b_i in the direction of $-c_i$. This contradicts Q being in $V(b_1, \ldots, b_k)$.

Thus we obtain that $M = V(b_1, \ldots, b_k)$. $\square$

The above result implies the following theorem.

Theorem 1. *A $k_{DIR}-CONV$ graph on n vertices can have at most n^k maximal cliques.*

Proof: By Proposition 1, every clique can be defined by a set $V(b_1, \ldots, b_k)$. Since there are at most n^k ways to choose $V(b_1, \ldots, b_k)$, it follows that there can be at most n^k cliques and thus at most n^k maximal cliques. $\square$

The above result can be somewhat strengthened, as the following theorem demonstrates. While this improvement still remains in $O(n^k)$, it does lower the bound implied by Proposition 1.

Theorem 2. *A $k_{DIR}-CONV$ graph can have at most $H_n^{(1-k)}$ maximal cliques, where $H_s^{(r)}$ is the generalized harmonic number with respect to s and r.*

Proof (Sketch): It is clear that not every one of the n^k combinations to choose $V(b_1, \ldots, b_k)$ is valid. Once P_1 through P_{i-1} are chosen, the only valid P_i is one that is at least as far as all previous P's in the direction of c_i. For simplicity, let us look only at the condition that it must be farther than P_1. Note that this is a less strict condition than it being farther than all previous P's, and as such will still result in a valid upper bound.

Once P_1 is chosen to be some polygon Q, for any $\mathbf{c}_i$, there are $d_i(Q)$ possible choices of P_i where $d_i(Q)$ is the number of polygons not strictly farther than Q in the direction $\mathbf{c}_i$. This means that if we choose each polygon Q_j as P_1, then for each Q_j the number of valid combinations of P_2 through P_k can be reduced from n^{k-1} to $\Pi_{i=2}^{k}(d_i(Q_j))$, bringing the total bound to $\Sigma_{j=1}^{n}\Pi_{i=2}^{k}(d_i(Q_j))$. This sum can be maximized by taking the mass product to be the highest it possibly can for $j = 1$; the second highest it possibly can for $j = 2$, etc. This occurs when, for $j = 1$, $d_i(Q_j) = n$ for all $i \geq 2$, for $j = 2$, $d_i(Q_j) = n - 1$ for all $i \geq 2$, etc. With this in mind, the upper bound $\Sigma_{j=1}^{n}\Pi_{i=2}^{k}(d_i(Q_j))$ simplifies to $\Sigma_{i=1}^{n} i^{k-1} = H_n^{(1-k)}$. Thus we obtained that the number of maximal cliques in a $k_{DIR} - CONV$ graph is upper-bounded by $H_n^{(1-k)}$.　　□

While the above result does not improve the asymptotic bound of Theorem 1, it is still a substantial refinement on n^k for an upper bound on the number of maximal cliques in a $k_{DIR} - CONV$ graph.

3.2　A Tight Bound When the Number of Directions is Fixed

Denote by $const_{DIR} - CONV$ the class of intersection graphs $k_{DIR} - CONV$ under the condition that the number of directions $k \geq 2$ is fixed. The following theorem shows that for this class the bound of Theorem 1 is tight.

Theorem 3. *A $const_{DIR} - CONV$ graph G on n vertices can have $\Theta(n^k)$ maximal cliques.*

Proof: We use an approach analogous to a proof in [18]. It is well-known (see [18, 21]) that for every $k \geq 2$, the class of intersection graphs $k - DIR$ is a proper subset of $k_{DIR} - CONV$, i.e., the relations $k - DIR \subseteq k_{DIR} - CONV$ and $k_{DIR} - CONV \nsubseteq k - DIR$ hold. Hence, it suffices to show that the statement is true for $k - DIR$ graphs.

Take a set S of n segments that satisfy the following requirements: (i) There are $\frac{n}{k}$ segments parallel to each of the k directions in L, and (ii) Every two nonparallel segments in S intersect.

Let G be the intersection graph of S. Then every maximal clique contains exactly one segment from each direction. Hence, the number of maximal cliques in G is $\left(\frac{n}{k}\right)^k = (1/k)^k n^k = \Omega(n^k)$ for any fixed integer k. This, coupled with the upper bound of Theorem 1, implies the stated result.　　□

3.3　Algorithmic Aspects

The results of the previous section do imply an $O(n^{k+2})$ algorithm for computing all maximal cliques in a $k_{DIR} - CONV$ graph. For this, one can first construct all $\mathbf{V}(b_1, \ldots, b_k)$ sets in $O(n^k)$ time. Then for each of these no more than n^k sets, one verifies in $O(n^2)$ time if the corresponding clique is maximal.

It is clear that the practical implementation of this general algorithm to work on any arbitrary input would be cumbersome. Nevertheless, this result provides a clear guide for designing an algorithm which would function similarly to those from [19] (for the case of homothetic triangles) and especially from [3] (for the case of quasi-homothetic trapezoids). In particular, this can also be implemented in-place (i.e. using only a constant amount of memory in addition to the input) analogously to the implementation in [3]. The only substantial change in methodology would be in defining which of two polygons is a better choice of b_i. In [3] it is done by comparing them in three directions, and then comparing their top edge as a tie breaker. Similarly, in the more general case considered here, one can compare two polygons' q_i sets for all i, and then as a tie breaker their q_i' sets for all i, all in the same cyclic manner demonstrated in [3]. From a practical standpoint, this result can be particularly useful for the widely studied cases of congruent, homothetic, or quasi-homothetic polygons. In all these, the parameter k is constant and thus the computation would feature polynomial time-complexity whose exponent is constant for all inputs. While any given input has a finite k, and therefore a polynomially computable solution, for cases such as random convex polygons, adding more polygons to the input is liable to define more directions.

Let us also mention that in [18] the authors remark that maximal cliques in a graph can be found by using the polynomial delay approach of [16]. Note that this would require an $O(n^3)$ delay between the generation of two subsequent maximal cliques, that would increase the overall computation time by a factor of n^3. Thus a computation based on Proposition 1 provides superior time complexity compared to the polynomial delay technique.

4 Pairs of Parallel Sides, Weak Sides, and Upper Bounds on the Number of Maximal Cliques in P_{hom} Graphs

In the considerations of this section we use the following technical notion from [18]. Given a convex polygon P, let s_1, s_2, and s_3 be three consecutive sides and l_1, l_2, and l_3 the corresponding straight lines containing them. Let H be the half-plane with boundary l_2 which does not contain P. The side s_2 is called *weak* if l_1 and l_3 do not intersect in H and there is no side parallel to s_2.

In [18] the authors prove the following.

Fact 2 *Given a convex polygon P with p sides, q pairs of parallel sides, and s weak sides, any P_{hom} graph contains at most $n^{2(p-q)-s}$ maximal cliques.*

While it is obvious that the bound n^k of the previous section is an improvement on the bound n^{2k} from [18], it is not immediately as clear how it relates to $n^{2(p-q)-s}$. The rest of the present section answers this question.

4.1 Parallel Sides and Weak Sides

It is observed in [18] that for a convex polygon P with p sides and q pairs of parallel sides, every P_{hom} graph is a $(p-q)_{DIR} - CONV$ graph. Then the n^{2k} upper bound for $k_{DIR} - CONV$ graphs implies an $n^{2(p-q)}$ bound for P_{hom} graphs. With some more detailed analysis this bound is reduced to $n^{2(p-q)-s}$.

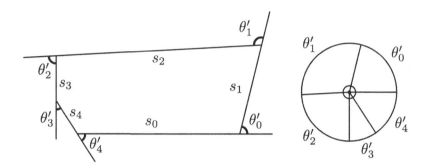

Fig. 3. *Left:* A polygon whose sides and corresponding external angles θ' are labelled. *Right:* All the θ' angles imposed on a circle to exemplify that they sum to 2π.

Next we determine the possible number of weak sides, its relation to the number of pairs of parallel sides, and the effect on the last upper bound when both pairs of parallel sides and weak sides are present in P.

Clearly, every triangle has three weak sides and every convex quadrilateral different from a parallelogram or trapezoid has exactly two weak sides. For arbitrary convex polygons with at least four sides we have the following facts.

Let P be a convex polygon with n sides $s_1, \ldots, s_n$. We define θ_i to be the angle between s_i and s_{i+1} where $0 < \theta_i < \pi$. Note that if $i = n$, then s_{i+1} and θ_{i+1} are s_1 and θ_1, and if $i = 1$, s_{i-1} and θ_{i-1} are s_n and θ_n, respectively (see Figure 3). The following is a plain fact.

Fact 3 *A side s_i is weak iff $\theta_{i-1} + \theta_i \le \pi$ and no other side is parallel to s_i.*

We define θ'_i to be the supplement of angle θ_i. Thus, θ'_i is the exterior angle corresponding to θ_i. The following is another well-known fact.

Fact 4 $\sum_{i=0}^{n} \theta'_i = 2\pi$.

We will show that a convex polygon can have no more than two weak sides. For this, we will use the following lemma.

Lemma 1. *A convex polygon with at least four sides can have no more than two consecutive weak sides.*

Proof: Let s_i, s_{i+1}, and s_{i+2} be three consecutive sides of P and assume they are all weak. Then we have that $\theta_{i-1} + \theta_i \le \pi$, $\theta_i + \theta_{i+1} \le \pi$, and $\theta_{i+1} + \theta_{i+2} \le \pi$. This implies that $\theta'_{i-1} + \theta'_i \ge \pi$ and $\theta'_{i+1} + \theta'_{i+2} \ge \pi$. Therefore, $\theta'_{i-1} + \theta'_i + \theta'_{i+1} + \theta'_{i+2} \ge 2\pi$. However, we know that $\sum_{i=1}^{n} \theta'_i = 2\pi$, so $\theta'_{i-1} + \theta'_i + \theta'_{i+1} + \theta'_{i+2} = 2\pi$. This means that s_{i-1}, s_i, s_{i+1}, and s_{i+2} are the only sides in P. Hence, s_{i+2} is the only side of P that is not adjacent to s_i. Since s_i and s_{i+2} are both weak, s_{i-1} and s_{i+1} must be parallel, which contradicts the assumption s_{i+1} being weak. Thus, no three consecutive sides can be weak, i.e., no more than two consecutive sides can be weak. $\square$

Proposition 2. *A convex polygon can have no more than two weak sides.*

Proof: Assume a convex polygon P has $n > 3$ sides, more than two of which are weak. By Lemma 1, P has no more than two consecutive weak sides, so we can distinguish between the following two cases:

Case 1: Any two weak sides of P are adjacent
Let s_i, s_j, s_k be weak sides of P, and let s_i and s_j be adjacent. Then $\theta'_{i-1} + \theta'_i \geq \pi$, $\theta'_i + \theta'_j \geq \pi$, and $\theta'_{k-1} + \theta'_k \geq \pi$.

Similar to the proof of Lemma 1, we obtain $\theta'_{i-1} + \theta'_i + \theta'_{k-1} + \theta'_k = 2\pi$, which implies that P is a quadrilateral. Then its three weak sides s_i, s_j, and s_k must be consecutive, which contradicts Lemma 1.

Case 2: No two weak sides of P are adjacent

Let s_i and s_j be weak sides of P. Then $\theta'_{i-1} + \theta'_i \geq \pi$ and $\theta'_{j-1} + \theta'_j \geq \pi$. Similar to above, we obtain $\theta'_{i-1} + \theta'_i + \theta'_{j-1} + \theta'_j = 2\pi$. Hence, there are only four angles in P, i.e., P is a quadrilateral. If more than two of its sides are weak, then there must be adjacent weak sides, which contradicts our assumption. Thus we obtain that a convex polygon can have no more than two weak sides.   □

4.2 Relationship between Parallel Sides and Weak Sides

For all that follows, see Figure 4 for reference and further explanation. The weak side definition and polygon's convexity imply that, if there is a pair of parallel sides, and if between one's endpoint and the other's endpoint there are two or more sides, they cannot be weak. It is also easy to see that, given a side s_i in a convex polygon, if side s_{i+1} is weak, then either side s_{i+2} is parallel to s_i or there is no side parallel to side s_i. Moreover, if s_{i+2} is not parallel to s_i, then any parallel sides after s_{i+2} would cause the polygon to be non-convex.

The above facts mean that if there are parallel sides, any weak side will connect them directly (i.e., an endpoint of the one would be linked by a side to an endpoint of the other). It follows that, given a single pair of parallel sides and assuming that there are sides (either one or two) that are weak, then there cannot be other parallel pairs of sides. This is because a weak side has to connect

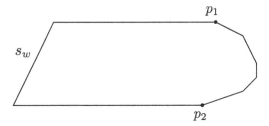

Fig. 4. Given that side s_w is weak, one can see that if its two adjacent sides are not parallel, then no side can be parallel to either of them. Furthermore, there cannot be a pair of parallel sides that are both to the right of points p_1 and p_2.

the given parallel sides directly, whereas by the convexity of polygon P there cannot be a pair of parallel sides that exist solely between two endpoints of the other parallel pair of sides.

The weak side that connects a parallel pair would therefore have to be a part of any second parallel pair, contradicting its being weak. Thus, we can conclude that if a convex polygon has weak sides, there is at most one pair of parallel sides, and if a convex polygon has a pair of parallel sides, the only possible weak sides are those which directly connect the two parallel sides.

4.3   Implication to the Bound on the Number of Maximal Cliques

Based on the preceding analysis we can conclude that there are relatively few convex polygons which have both parallel pairs and weak sides. More specifically, these polygons can have only one parallel pair, and either of its possible weak sides would have to connect the parallel sides directly. This means that the upper bound $n^{2(p-q)-s}$ on the number of maximal cliques applies only to trapezoids or trapezoid-like polygons (such as the one depicted in Figure 4) where one of the non-parallel sides is replaced by a chain of sides.

We can see then that this bound is an improvement of a power of two for trapezoids and of a power of one for trapezoid-like polygons on the bound $n^{2(p-q)}$, while for all other classes of polygons either $q = 0$ or $s = 0$ or both, and if $q = 0$ then $s \leq 2$, which at best reduces the power by two.

With the above analysis it is now not hard to compare the bound n^k from the previous section and the bound $n^{2(p-q)-s}$ from [18]. One can see that $p - q$ is exactly k. Since $s \leq 2$, it follows that in order for $n^{2(p-q)-s}$ to be a tighter upper bound than n^k, it must be the case that $2k - 2 < k$. The latter can only be true if $k < 2$, which is not possible for a convex polygon. One can also see that if $k \geq 3$, n^k is a strictly tighter bound. The only exceptions to this are the triangles, for which $k = 3$. However, $s = 3$ as well, so that even in this extreme case the two bounds appear to be equal.

As a last remark we note that Fact 2 implies an upper bound of n^4 for the number of maximal cliques when applied to trapezoids [18]. In a recent work [3] we showed that this bound can be reduced to n^3, which appears to be tight.

5   Concluding Remarks

In this paper we presented new upper bounds on the number of maximal cliques in $k_{DIR} - CONV$ graphs. For a graph representation with n polygons we proved an upper bound of n^k which is a considerable improvement on the known n^{2k} bound from [18]. We also showed that this bound is tight for any fixed k. An open question is whether the same holds for a variable k.

The proof of the main result is constructive and provides a guide to devising a space-efficient (in-place) algorithm for computing the maximal cliques in a graph from the considered class. Its implementation for some special cases of practical importance is seen as an important future task. We believe that the construction of the sets $\mathbf{V}(b_1, \ldots, b_k)$ in $\mathbb{R}^2$ can be generalized to convex polytopes in higher dimensions.

Acknowledgements. The authors thank the three anonymous referees for their useful remarks and suggestions.

This work was done in the framework of the undergraduate research program "URGE to Compute" sponsored by NSF grants No 0802964 and No 0802994.

References

1. Ambühl, C., Wagner, U.: The clique problem in intersection graphs of ellipses and triangles. Theory Comput. Syst. 38, 279–292 (2005)
2. Ballard, D.H., Brown, M.: Computer Vision. Prentice-Hall, Englewood Cliffs (1982)
3. Brimkov, V.E., Kafer, S., Szczepankiewicz, M., Terhaar, J.: Maximal cliques in intersection graphs of quasi-homothetic trapezoids. In: Proc. MCURCSM 2013, Ohio, 10 p. (2013)
4. Cabello, S., Cardinal, J., Langerman, S.: The clique problem in ray intersection graphs. In: Epstein, L., Ferragina, P. (eds.) ESA 2012. LNCS, vol. 7501, pp. 241–252. Springer, Heidelberg (2012)
5. Čulík, K.: Applications of graph theory to mathematical logic and linguistics. In: Proc. Sympos. "Theory of Graphs and its Applications" (Smolenice, 1963), pp. 13–20. Publ. House Czechoslovak Acad. Sci., Prague (1964)
6. Erdős, P, Goodman, A.W., Pósa, L.: The representation of a graph by set intersections. Canad. J. Math. 18, 106–112 (1966)
7. Evako, A.V.: Topological properties of the intersection graph of covers of n-dimensional surfaces. Discrete Mathematics 147(1-3), 107–120 (1995)
8. Felsner, S., Müller, R., Wernisch, L.: Trapezoid graphs and generalizations, geometry and algorithms. Discrete Applied Mathematics 74, 13–32 (1993)
9. Fish, A., Stapleton, G.: Formal issues in languages based on closed curves. In: Proc. Distributed Multimedia Systems, pp. 161–167 (2006)
10. Gardiner, E.J., Artymiuk, P.J., Willett, P.: Clique-detection algorithms for matching three-dimensional molecular structures. J. Molecular Graph Modelling 15(4), 245–253 (1997)
11. Golumbic, M.: Algorithmic Graph Theory and Perfect Graphs. Acad. Press (1980)
12. Heinzle, F., Ander, K.H., Sester, M.: Graph based approaches for recognition of patterns and implicit information in road networks. In: Proc. 22nd International Cartographic Conference, A Coruna (2005)
13. Imai, H., Asano, T.: Finding the connected components and a maximum clique of an intersection graph of rectangles in the plane. Journal of Algorithms 4, 300–323 (1983)
14. Ion, A., Carreira, J., Sminchisescu, C.: Image segmentation by figure-ground composition into maximal cliques. In: Proc. 13th International Conference on Computer Vision, Barcelona, pp. 2110–2117 (2011)
15. Jacobson, M.S., Morris, F.R., Scheinermann, E.R.: General results on tolerance intersection graphs. J. Graph Theory 15, 573–577 (1991)
16. Johnson, D.S., Yannakakis, M., Papadimitriou, C.H.: On generating all maximal independent sets. Information Processing Letters 27, 119–123 (1988)
17. Klette, R., Rosenfeld, A.: Digital Geometry. Geometric Methods for Digital Picture Analysis. Morgan Kaufmann, San Francisco (2004)

18. Junosza-Szaniawski, K., Kratochvíl, J., Pergel, M., Rzążewski, P.: Beyond homothetic polygons: Recognition and maximum clique. In: Chao, K.-M., Hsu, T.-s., Lee, D.-T. (eds.) ISAAC 2012. LNCS, vol. 7676, pp. 619–628. Springer, Heidelberg (2012)
19. Kaufmann, M., Kratochvíl, J., Lehmann, K., Subramanian, A.: Max-tolerance graphs as intersection graphs: cliques, cycles, and recognition. In: Proc. SODA 2006, pp. 832–841 (2006)
20. Kratochvíl, J., Kuběna, A.: On intersection representations of co-planar graphs. Discrete Mathematics 178, 251–255 (1998)
21. Kratochvíl, J., Matoušek, J.: Intersection graphs of segments. J. Combinatorial Theory Ser. B 62, 289–315 (1994)
22. Kratochvíl, J., Nešetřil, J.: Independent set and clique problems in intersection-defined classes of graphs. Comm. Math. Uni. Car. 31, 85–93 (1990)
23. Kratochvíl, J., Pergel, M.: Intersection graphs of homothetic polygons. Electronic Notes in Discr. Math. 31, 277–280 (2008)
24. McKee, T.A., McMorris, F.R.: Topics in Intersection Graph Theory. SIAM Monographs on Discrete Mathematics and Applications, vol. 2. SIAM, Philadelphia (1999)
25. Szpilrajn-Marczewski, E.: Sur deux propriétés des classes d'ensembles. Fund. Math. 33, 303–307 (1945)
26. Nakamura, H., Masatake, H., Mamoru, H.: Robust computation of intersection graph between two solids. Graphical Models 16(3), C79–C88 (1997)
27. Nandy, S.C., Bhattacharya, B.B.: A unified algorithm for finding maximum and minimum object enclosing rectangles and cuboids. Computers Math. Applic. 29(8), 45–61 (1995)
28. Paget, R., Longsta, D.: Extracting the cliques from a neighbourhood system. IEE Proc. Vision Image and Signal Processing 144(3), 168–170 (1997)
29. Simonetto, P., Auber, D.: An heuristic for the construction of intersection graphs. In: Proc. 13th International Conference on Information Visualisation, pp. 673–678 (2009)
30. Simonetto, P., Auber, D.: Visualise undrawable Euler diagrams. In: Proc. 12th IEEE International Conference on Information Visualisation, pp. 594–599 (2008)
31. Tian, J., Tinghua, A., Xiaobin, J.: Graph based recognition of grid pattern in street networks. In: The International Archives of the Photogrammetry, Remote Sensing and Spatial Information Sciences, Advances in Spatial Data Handling and GIS. Lecture Notes in Geoinformation and Cartography, Part II, vol. 38, pp. 129–143 (2012)
32. Vairinhos, V.M., Lobo, V., Galindo, M.P.: Intersection graph-based representation of contingency tables, http://www.isegi.unl.pt/docentes/vlobo/Publicacoes/3_17_lobo08_DAIG_conting_tables.pdf
33. Müller, T., van Leeuven, E.J., van Leeuven, J.: Integer representations of convex polygons intersection graphs. In: Symposium on Computational Geometry, pp. 300–307 (2011)
34. van Leeuwen, E.J., van Leeuwen, J.: Convex polygon intersection graphs. In: Brandes, U., Cornelsen, S. (eds.) GD 2010. LNCS, vol. 6502, pp. 377–388. Springer, Heidelberg (2011)
35. Verroust, A., Viaud, M.-L.: Ensuring the drawability of extended euler diagrams for up to 8 sets. In: Blackwell, A.F., Marriott, K., Shimojima, A. (eds.) Diagrams 2004. LNCS (LNAI), vol. 2980, pp. 128–141. Springer, Heidelberg (2004)
36. Wang, X., Bai, X., Yang, X., Wenyu, L., Latecki, L.J.: Maximal cliques that satisfy hard constraints with application to deformable object model learning. In: Advances in Neural Information Processing Systems, vol. 24, pp. 864–872 (2011)

Weighted Distances on a Triangular Grid

Benedek Nagy

Department of Mathematics, Faculty of Arts and Sciences
Eastern Mediterranean University,
Famagusta, North Cyprus, Mersin-10, Turkey
Department of Computer Science, Faculty of Informatics
University of Debrecen
PO box 12, 4010 Debrecen, Hungary
nbenedek@inf.unideb.hu

Abstract. In this paper we introduce weighted distances on a triangular grid. Three types of neighborhood relations are used on the grid, and therefore three weights are used to define a distance function. Some properties of the weighted distances, including metrical properties are discussed. We also give algorithms that compute the weighted distance of any point-pair on a triangular grid. Formulae for computing the distance are also given. Therefore the introduced new distance functions are ready for application in image processing and other fields.

Keywords: Triangular grid, Digital distances, Shortest paths, Digital metrics, Weighted distances, Chamfer distances, Distance map.

1 Introduction – Digital Distances

In a discrete space, which is usually used in image processing and computer graphics, there are some phenomena which do not occur in the Euclidean plane, and vice versa. For instance, there is a point (moreover there are infinitely many distinct points) between any two distinct points of the Euclidean space. Opposite to this fact, there are neighbor points (pixels) in digital spaces. The points of a discrete grid having distance r from a given point of the grid (e.g., the Origin) do not form a circle (in the usual sense), but usually they form a small finite set that is not connected in any sense. Therefore digital distances can be defined and used in various applications instead of the Euclidean distance [6]. The points of the regular grids can be addressed by integer coordinate values. In square (cubic and its higher dimensional versions) grid usually the Cartesian coordinate system is used. The pixels of the hexagonal grid can be addressed with two integers [7], but there is a more elegant solution using three coordinate values whose sum is zero reflecting the symmetry of the grid [5,8]. For the triangular grid three coordinate values can effectively be used which are not independent [9]. The theory of digital distances has begun and so the field digital geometry was born by a paper of Rosenfeld and Pfaltz. In [16] they introduced two types of neighborhood relations on the square grid (Manhattan/cityblock and chessboard) and distances based on them. The digital distances of two points based

R.P. Barneva, V.E. Brimkov, and J. Šlapal (Eds.): IWCIA 2014, LNCS 8466, pp. 37–50, 2014.

on a neighborhood relation gives the length of a shortest path connecting the two points where in each step we can move to a (given type) neighborhood point of the actual point. However both cityblock and chessboard distances are very rough approximations of the Euclidean distance: their rotational dependency is very large (in both cases the Euclidean distance of two points that have the same digital distance from the Origin can differ by a factor $\sqrt{2}$). Therefore some other digital distances were introduced. There are three well-known ways to reduce the rotational dependency of a digital distance function: the first method is to allow to use various types of neighbors at various steps in the paths (already Rosenfeld and Pfaltz gave an idea to use the two types of neighborhood alternating way in a path). In this way the digital distances based on neighborhood sequences were introduced and further developed in [20,1,2,12]. The second usual way to reduce the rotational dependency of a digital distance is by using various weights for various steps in a path, and then, computing the distance as the least cost of the weighted paths between the points [18,19,4]. We note here that weighted distances are also called chamfer distances. The third way could be to use a "non-traditional" grid instead of the square grid. Both the hexagonal and the triangular grids have some nice properties, e.g., rotations by $\frac{\pi}{3}$ transform the grid to itself (for the square grid rotations by $\frac{\pi}{2}$ are needed). They are dual grids of each other (in graph theoretic sense). Consequently, it can easily be seen that both of these grids gave better distance function based only on a given neighborhood than the cityblock and the chessboard distances [10]. In some cases one may mix these methods, e.g., there are weighted distances based on neighborhood sequence on the square grid [15]. The theory of distances based on neighborhood sequences on the triangular grid is also well developed [9,11,14]. However the theory of weighted distances for the triangular grid was a missing point in this field and therefore this paper starts to fill this gap by the basic definitions and basic results.

2   Description of the Triangular Grid

Similarly to the hexagonal grid [5,8], each pixel of the triangular grid can also be described by a unique coordinate-triplet [9,14] as we recall in Figure 1. Note that each pixel of the triangular grid is a triangle and we also call it as a point of the grid. There are two orientations of the used triangles: there are triangles of shape $\triangle$ and there are triangles of shape ∇. The sum of the coordinate values that address a point is zero or one depending the orientation (shape) of the triangle. The origin, the triangle having coordinate triplet $(0, 0, 0)$, and the directions of the axes can be seen on the figure. At every time when we step from a triangle to another one crossing their common side, the step is done parallel to one of the axes. If the step is in the direction of an axis, then the respective coordinate value is increased by 1, while in case the step is in opposite direction to an axis, the respective coordinate value is decreased by 1. In this way every triangle gets a unique coordinate triplet ([13]). However the three values are not independent (we are in a two dimensional space, i.e., plane); their sum is either 0 or 1. The

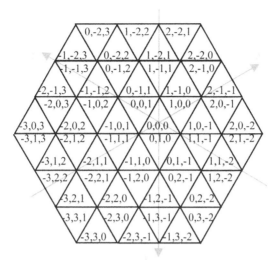

Fig. 1. Coordinate axes and coordinate system for the triangular grid

points having 0-sum coordinate values are called even, the points with 1-sum are called odd. One of the first things to notice about the grid is that the points are of two types (they have two types of parities); the even points have shape △, while the odd points have an opposite orientation. In this way the coordinate system reflect correctly the two types of points.

There are three types of (commonly used) neighborhood on this grid [3]. Two points are 1-neighbors if they share a side. Two triangles are strict 2-neighbors if they have a common 1-neighbor triangle. Two pixels are 3-neighbors if they share at least a point on their boundaries (e.g., a corner point). In Figure 2 a point and its twelve neighbors are shown. Using the coordinate triplets one can give the neighborhood relations in the following formal form. The points p and q of the triangular grid are
- m-neighbors ($m = 1, 2, 3$), if
  (1) $|p(i) - q(i)| \leq 1$, for $i = 1, 2, 3$,   and
  (2) $|p(1) - q(1)| + |p(2) - q(2)| + |p(3) - q(3)| \leq m$.
- strict m-neighbors, if there is an equality in (2) for the value of m.

The neighborhood relation is reflexive (i.e., each triangle is a 1-, 2-, and a 3-neighbor to itself). Two points are 1-neighbors if they are at most 1-neighbors in strict sense. Two points are 2-neighbors if they are strict 2-neighbors or 1-neighbors. Two pixels are 3-neighbors if they are strict 3-neighbors or 2-neighbors. We use three types of neighbors, as they can be seen on Figure 2 (the number written in the triangles indicate the value m for which the given triangle is a strict m-neighbor of the central triangle marked by 'O').

Observe that the 1-neighbors of the neighbors are exactly the 2-neighbors of the original point. The 2-neighbors of the 3-neighbors of a point are exactly the 3-neighbors of its 2-neighbors (moreover the 3-neighbors of the 3-neighbors

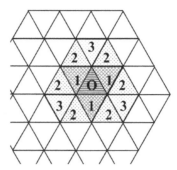

Fig. 2. Types of neighborhood relations on the triangular grid

form the same set of points). But the 1-neighbors of the 3-neighbors and the 3-neighbors of a 1-neighbors form different sets of points (they are incomparable by inclusion), and both of these sets are strictly included in the 2-neighbors of the 2-neighbors.

The set of points with a fixed coordinate value form a lane, e.g., $\{(x, y, z)|x + y + z \in \{0, 1\}, x = 2\}$. If two points share a coordinate value, then they are in a common lane (according to this shared coordinate value). If there is no coordinate value that is shared by the two points, then the points can be connected by two lanes having angle $\frac{2}{3}\pi$ between them, as it can be seen in Figure 3.

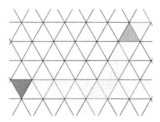

Fig. 3. Example for lanes, as they connect triangle pixels

Note, that the triangular grid is not a lattice, i.e., there are vectors which map some points to grid-points and some points to not grid-points. For instance, the vector $(1, 0, 0)$ maps the even points to odd points, but there are no images of the odd points in the grid. Only vectors with zero-sum values map the grid to itself [13].

3   Definition of Weighted Distances

We recall shortly, that on the square grid there are only two types of usual neighborhood relations that are widely used: the cityblock and the chessboard neighborhood. The simplest weighted distances allow to step to a cityblock neighbor by changing only 1 coordinate value by ± 1 with weight α and allows to step to a diagonal neighbor by changing both the coordinate values by ± 1 with weight β. There are weighted distances using larger neighborhood and also in higher dimensional grids [18,19,4]. In this paper, we consider the usual three types of neighborhood on the triangular grid.

Let the weight of a step to a 1-neighbor be α, of a step to a strict 2-neighbor be β and of a step to a strict 3-neighbor be γ. As natural, we assume that $0 < \alpha \leq \beta \leq \gamma$.

Then, according to the possible types of neighbors, we can define the weighted distance of any two points of the triangular grid.

Let $p = (p(1), p(2), p(3))$ and $q = (q(1), q(2), q(3))$ be two points of the triangular grid. A finite point sequence of points of the form $p = p_0, p_1, \ldots, p_m = q$, where p_{i-1}, p_i are 3-neighbor points for $1 \leq i \leq m$, is called a path from p to q. A path can be seen as consecutive steps to neighbor points. Then the cost (i.e., the weight) of the path is the sum of the weights of its steps: $\alpha |\{(p_i, p_{i+1}) \mid p_i$ and p_{i+1} are 1-neighbors, $i \in \{0, \ldots, m-1\}\}| + \beta |\{(p_i, p_{i+1}) \mid p_i$ and p_{i+1} are strict 2-neighbors, $i \in \{0, \ldots, m-1\}\}| + \gamma |\{(p_i, p_{i+1}) \mid p_i$ and p_{i+1} are strict 3-neighbors, $i \in \{0, \ldots, m-1\}\}|$. Finally, let the weighted distance $d(p, q; \alpha, \beta, \gamma)$ of p and q by the weights α, β, γ be the cost of a/the minimal weighted paths(s) between p and q.

Now we introduce some further technical definitions that are used through the paper. The difference $w_{p,q} = (w(1), w(2), w(3))$ of two points p and q is defined by: $w(i) = q(i) - p(i)$. If $w(1) + w(2) + w(3) = 0$, then the parity of w is even, else it is odd. Let $v_{p,q}$ be the sorted difference of the points, i.e., we order the values of $w_{p,q}$ in a non-increasing way by their absolute values. To have a uniquely determined $v_{p,q}$ if equal values occur, then we prefer $v(1) > 0$ and $v(1)v(2) < 0$. In obvious cases we omit the indices of $w_{p,q}$ and $v_{p,q}$ and use simply w and v instead of $w_{p,q}$ and $v_{p,q}$, respectively.

4 Algorithm for a Minimum Weighted Path and Computing the Distances

There are various paths with various sums of weights that can be found between any two points. In Figure 4 we display several possible solutions for a minimal weight path (depending on the actually used weights). One could obtain the values of the figure by a breadth-first-search like algorithm. When the weights α, β and γ are known the optimal-search, i.e., the Dijkstra algorithm can also be used (keeping the best sum of weights, i.e., the actual best found distance value, and finally the distance itself for every points from that the algorithm is already continued) [17]. In a very similar way one may construct the distance map of a binary image, by measuring the weighted distance of the background points from the closest foreground points. Here we do not detail this algorithm, we concentrate on the theoretical background of the weighted distances on the triangular grid.

However depending on the actual ratios and values of the weights α, β, γ, one can compute a minimum weighted paths, and so the distance of arbitrary

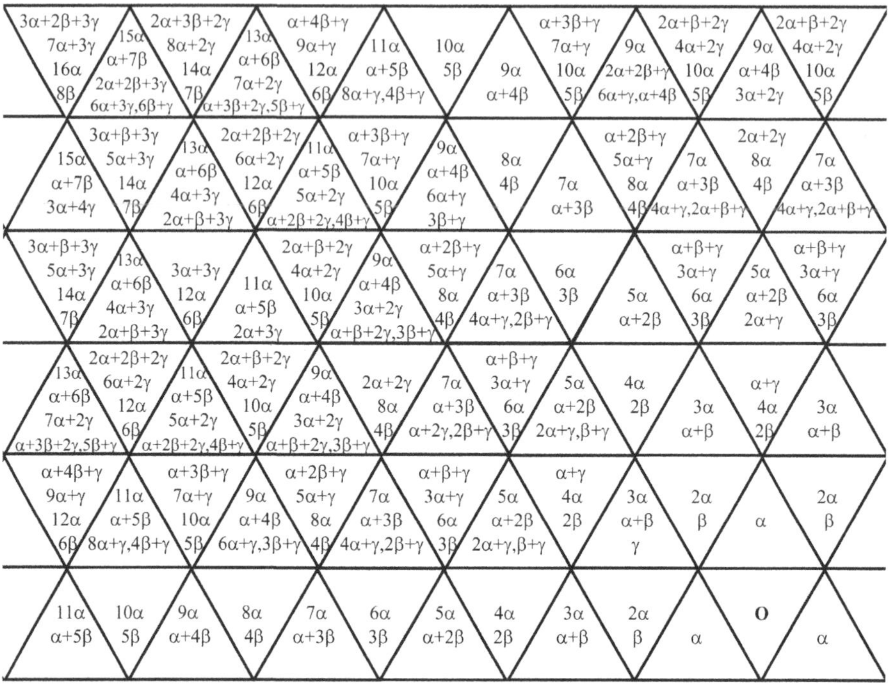

Fig. 4. The possible minimal weights of points reached in a path from the point **O**

two points of the grid in a much faster way than the previously shown method. In the next part, using a combinatorial approach, we give methods for these computations for each possible case.

As we already mentioned, we use the natural condition $0 < \alpha \le \beta \le \gamma$ for the used weight values. We know that with a 1-step (by weight α) only 1 of the coordinates changes by ± 1; with a strict 2-step (by weight β) exactly 2 of the coordinates change by $+1$ and -1, respectively; and with a strict 3-step (by weight γ) every coordinate changes by ± 1. Therefore it is important to measure the relative weight of a step, that is the cost of the change of a coordinate value by ± 1. These relative weights give the first separation of the possible cases.

4.1   Case $2\alpha \le \beta$ and $3\alpha \le \gamma$

Let us consider first the case when α has the lowest relative weight value. (An example: $\alpha = 1, \beta = 3, \gamma = 4$.) Every strict 2-step can be substituted by two consecutive 1-steps; and every strict 3-step can be substituted by three consecutive 1-steps. Moreover by substituting a 2-step by 1-steps the cost of the path is not increasing since $2\alpha \le \beta$ (and the cost strictly decreasing if $2\alpha < \beta$). Similarly, by substituting a 3-step by 1-steps the cost of the path is not increasing since $3\alpha \le \gamma$ (and the cost strictly decreasing if $3\alpha < \gamma$). Therefore the cost of any path between any two points is not increasing by changing all its steps to 1-steps. Therefore paths build up by only 1-steps can have minimal cost. To have a minimal cost path we need path(s) with 1-steps that have minimal number of steps.

Actually, a minimal path can be constructed in the following way: let us start from point p and compute the vector $w_{p,q}$. Then, in a minimal cost path, in every step the absolute value of a coordinate difference is decreasing by 1. Thus

$$d(p, q; \alpha, \beta, \gamma) = \alpha(|w(1)| + |w(2)| + |w(3)|).$$

4.2   Case $2\alpha > \beta$ and $3\alpha \le \gamma$

In this case the strict 2-steps, β has the lowest relative weight. (An example: $\alpha = 2, \beta = 3, \gamma = 7$.) It is easy to see that usage of 3-steps is not efficient: three 1-steps give a value that is not larger. Since every two consecutive 1-steps can be replaced by a 2-step (and vice versa), optimal paths can be constructed in the following way:

If there are at least two non-zero values in $w_{p,q}$, then a 2-step decreases the absolute values of two non-zero coordinates of w, otherwise a 1-step is needed to reach the end point q.

Thus, if the parity of p and q are the same, then

$$d(p, q; \alpha, \beta, \gamma) = \beta \frac{|w(1)| + |w(2)| + |w(3)|}{2};$$

and if the parities of p and q differ, then

$$d(p, q; \alpha, \beta, \gamma) = \alpha + \beta \frac{|w(1)| + |w(2)| + |w(3)| - 1}{2}.$$

4.3   Case $2\alpha > \beta$ and $3\alpha > \gamma$, but $\alpha + \beta \le \gamma$

In this case 2-steps are preferred over 1-steps again. The 3-steps has lower relative weights than the 1-steps, but a 3-step can be changed to a 2-step and a 1-step without increasing the cost of the path, therefore there are optimal paths without 3-steps. (An example: $\alpha = 4, \beta = 6, \gamma = 11$.)

In this way, this case is similar to the previous one. Optimal path(s) can be constructed using 2-steps (and maybe a 1-step if the parities of the two points p and q differ). Thus, the distance is

$$d(p, q; \alpha, \beta, \gamma) = \beta \frac{|w(1)| + |w(2)| + |w(3)|}{2}$$

if p and q have the same parities; and

$$d(p, q; \alpha, \beta, \gamma) = \alpha + \beta \frac{|w(1)| + |w(2)| + |w(3)| - 1}{2}$$

otherwise.

4.4   Case $2\alpha > \beta$ and $3\alpha > \gamma$, but $\alpha + \beta > \gamma$

Now we are continuing with the cases where 3-steps can be worthwhile. In this case 2-steps have lower relative weights than 1-steps. Actually there are two subcases of the present case.

4.4.1   Subcase $\gamma + \alpha > 2\beta$

Since $\alpha + \beta > \gamma$, $2\alpha + \beta > \gamma + \alpha > 2\beta$. (An example: $\alpha = 5, \beta = 7, \gamma = 10$.) In an optimal path there can be a 3-step, but only once. (To effectively use a 3-step again, one need to change the parity of the point and that can only be by a 1-step (in a given direction after a 3-step is made). Since a 3-step and a 1-step together can be substituted by two 2-steps with a lower cost optimal path do not contain both 3-steps and 1-steps.)

There are some cases according to the respective places and parities of the points.

- If p and q have the same parities, then an optimal path connects them with only 2-steps reducing the absolute values of two coordinates of the difference vector. (This path can be on the common lane of the points p and q, if there is such a lane; or it can be done by two lanes connecting the points with angle $\frac{2}{3}\pi$.) In this way

$$d(p, q; \alpha, \beta, \gamma) = \beta \frac{|w(1)| + |w(2)| + |w(3)|}{2}.$$

- If p and q have different parities, then two possibilities may occur.

  – p is even and w contains a negative and two positive values, or p is odd an w contains a positive and two negative values, then an optimal path can be constructed as follows:

First we have 2-steps in a lane by reducing one of the positive/negative values (in case of p even/odd, respectively) to have absolute value 1. Then a 3-step can be done by reaching another lane (in which q is located), and so q can be reached by 2-steps reducing the other two coordinate differences to 0. Thus,

$$d(p, q; \alpha, \beta, \gamma) = \beta \frac{|w(1)| + |w(2)| + |w(3)| - 3}{2} + \gamma.$$

  – otherwise, a minimal cost path contains 2-steps and a 1-step on the common lane of p and q, if they are on a common lane; or the points can be connected by two lanes with angle $\frac{2}{3}\pi$ by 2-steps and a 1-step.

In this way

$$d(p, q; \alpha, \beta, \gamma) = \alpha + \beta \frac{|w(1)| + |w(2)| + |w(3)| - 1}{2}.$$

4.4.2   Subcase $\gamma + \alpha \leq 2\beta$

In this subcase if a 3-step and a 1-step can be applied (in any order) it is worth to apply them instead of two 2-steps. (An example: $\alpha = 4, \beta = 7, \gamma = 9$.)

Actually, if the directions of two consecutive 2-steps are not the same, then they can be replaced by a 3-step and a 1-step. Two 2-steps in the same direction, i.e., stepping on the same lane, cannot be replaced by a 3-step and a 1-step. An optimal path from p to q can be constructed as follows:

ALGORITHM 1:

input : the points p and q and the weights α, β, γ with the condition $\gamma + \alpha \leq 2\beta$
output : an optimal path Π from p to q as a sequence of points and its cost
let cost=0 and $\Pi = (p)$
while (w has no zero value) do
    if (p is even and w has a negative and two positive values
    or p is odd and w has a positive and two negative values)
      then let the new p be obtained from p by a 3-step
        decreasing the absolute value of each component of w,
        let cost=cost+γ
      else let the new p obtained from p by a 1-step
        decreasing the absolute value of a component of w,
        let cost=cost+α
    concatenate p to Π
end while
while ($p \neq q$) do
    let p be obtained from p by decreasing the absolute value of
      every nonzero component of w,
    concatenate p to Π
    if two components were changed then let cost=cost+β
      else let cost=cost+α
end while

After the execution of the algorithm cost is the distance of the original input points p and q.

For the formula to compute the distance, we can distinguish the following cases.

- if p and q have different parities and
  – p is even and w has a negative and two positive values or p is odd and w has a positive and two negative values, then

$$d(p, q; \alpha, \beta, \gamma) = \alpha(|v(3)| - 1) + \beta \frac{|v(1)| + |v(2)| - 3|v(3)| + 1}{2} + \gamma |v(3)|.$$

  – p is odd and w has a negative and two positive values or p is even and w has a positive and two negative values, then

$$d(p, q; \alpha, \beta, \gamma) = \alpha(|v(3)| + 1) + \beta \frac{|v(1)| + |v(2)| - 3|v(3)| - 1}{2} + \gamma |v(3)|.$$

  – if w has zero value, then

$$d(p, q; \alpha, \beta, \gamma) = \alpha + \beta \frac{|w(1)| + |w(2)| + |w(3)| - 1}{2}.$$

- if p and q has the same parity (i.e., their difference vector is even), then

$$d(p, q; \alpha, \beta, \gamma) = \alpha|v(3)| + \beta\frac{|v(1)| + |v(2)| - 3|v(3)|}{2} + \gamma|v(3)|.$$

4.5   Case $2\alpha \leq \beta$ and $3\alpha > \gamma$

In this case the 3-steps have the lowest relative weights and therefore we should use them as many as possible. On the other side, the 1-steps have lower (or equal) relative weights than the 2-steps and since each 2-step can be replaced by two 1-steps, therefore there are optimal paths without any 2-steps. (An example: $\alpha = 3, \beta = 7, \gamma = 8$.)

Optimal path can be constructed between p and q as follows:

ALGORITHM 2:
input: the points p and q and the weights α, β, γ with the conditions $2\alpha \leq \beta$ and $3\alpha > \gamma$
output: an optimal path Π from p to q and its cost
let cost=0 and $\Pi = (p)$
while (w has no zero value) do
  if (p is even and w has a negative and two positive values
    or p is odd and w has a positive and two negative values)
      then let the new p be obtained from p by a 3-step
        decreasing the absolute value of each component of w,
        let cost=cost+γ
      else let the new p obtained from p by a 1-step
        decreasing the absolute value of a component of w,
        let cost=cost+α
  concatenate p to Π
end while
while ($p \neq q$) do
  let p be obtained from p by decreasing the absolute value of
    a nonzero component of w,
  let cost=cost+α
  concatenate p to Π
end while

After the execution of the algorithm cost is the distance of the original input points p and q.

For the formula to compute the distance, we have

$$d(p, q; \alpha, \beta, \gamma) = \alpha(|v(1)| + |v(2)| - 2|v(3)|) + \gamma|v(3)|.$$

Table 1 is recapitulating the cases. These conditions are the bases to know what types of steps are used in a minimal weighted path. As we could see, in some cases, a path with minimum weight can be obtained, and thus, the weighted distance can be computed in a very simple way using steps to 1- and/or 2-neighbors on some specific lanes. Since the triangular grid is not a lattice, in some

Table 1. Summary of the possible cases of the weighted distance depending on the weights. Case i is described in Subsection 4.i.

case	$2\alpha \leq \beta$	$3\alpha \leq \gamma$	$\alpha + \beta \leq \gamma$	$\gamma + \alpha > 2\beta$
case 1	Yes	Yes		
case 2	No	Yes		
case 3	No	No	Yes	
case 4.1	No	No	No	Yes
case 4.2	No	No	No	No
case 5	Yes	No		

cases (when steps to strict 3-neighbors are more advantageous than smaller steps instead) the description of the computation seems to be more complex. In these cases steps to strict 3-neighbors and to 1-neighbors are used in an alternating way till the direction is appropriate; after that steps to 1- and/or strict 2-neighbors are used (depending also on the case). Our algorithms are greedy and their time complexity is linear on the difference of the points, i.e., on $|w(1)|+|w(2)|+|w(3)|$. However if only the weighted distance is required, we can compute it directly by the appropriate formula. In Figure 4, for the given region of the grid, every formula (i.e., value for every possible case) is presented. They can be computed by the actual weights and the smallest value obtained in a triangle gives its weighted distance from the triangle marked by 'O'. However, if the actual values of the weights are known, the distance can be directly computed by using the formula described at the respective case.

5   Properties of the Distances

Now we are detailing some properties of the weighted-distances on the triangular grid. Let us start with the metric properties. To have a metrical distance three properties must be fulfilled: positive definiteness, symmetry and triangular inequality.

Theorem 1. *For any positive values of α, β, γ the weighted distance function is a metric.*

Proof. Let α, β and γ be arbitrary positive values.

First we show that the positive definiteness is fulfilled: it is clear that for any point p, $d(p, p; \alpha, \beta, \gamma) = 0$ and also, if $d(p, q; \alpha, \beta, \gamma) = 0$, then $p = q$. Moreover, $d(p, q; \alpha, \beta, \gamma) \geq 0$ for any point pair p, q of the grid.

Since every shortest path is a shortest path having its points in reverse order, we can conclude that these distances are symmetric:

$$d(p, q; \alpha, \beta, \gamma) = d(q, p; \alpha, \beta, \gamma)$$

for every pair of points p and q.

Finally, let us consider the triangular inequality. Contrary, assume that there are weights α, β, γ and three points p, q and r such that

$$d(p, r; \alpha, \beta, \gamma) + d(r, q; \alpha, \beta, \gamma) < d(p, q; \alpha, \beta, \gamma).$$

Then there is a shortest path from p to r with length $d(p, r; \alpha, \beta, \gamma)$ and there is a shortest path from r to q with length $d(r, q; \alpha, \beta, \gamma)$. Thus there is a path between p and q obtained by the concatenation of the above two shortest paths with length $d(p, r; \alpha, \beta, \gamma) + d(r, q; \alpha, \beta, \gamma)$. But then, the distance of p and q cannot be larger than the length of this path between them. There is a contradiction that proves the triangular inequality of these distances.

Thus every weighted distance on the triangular grid has the metric properties.

$\square$

The previous theorem is interesting by the mirror of the fact that distances based on neighborhood sequences can be non metrical ones ([9,14]). The triangular inequality can easily been violated. Moreover on the triangular grid there are distances based on neighborhood sequences that are non-symmetric. Using distances based on neighborhood sequences, a sequence B of possible neighborhoods give restrictions on the steps in a (shortest) B-path (and unit weight used in each step). To see these non-metrical distances let us consider the neighborhood sequence $B = (3, 1, 1, 1, 1, \ldots)$: it allows to step to any 3-neighbor (allowing 1- and 2-neighbors also) of the starting point in the first step and then the path must be continued only by steps to a 1-neighbor of the previous point. Let $p = (0, 0, 0)$ and $q = (1, -2, 1)$. Then one can obtain a shortest B-path from p to q with two steps (having intermediate point $(1, -1, 1)$), and thus $d(p, q; B) = 2$. However there is no B-path from q to p with length less than three. Therefore $3 = d(q, p; B) \neq d(p, q; B)$. Hence the B-distance based on this neighborhood sequence is not symmetric for any pairs of points. Moreover, it does not satisfy the triangular inequality: let $r = (-1, 2, -1)$, then it can be seen that $d(r, p; B) = 2$ and $d(r, q; B) = 6$. Since $4 = d(r, p; B) + d(p, q; B) < d(r, q; B) = 6$, the triangular inequality is violated by this B-distance. (One may easily follow these steps and distances on Figure 1.)

Theorem 2. *The weighted distances are translation invariant for vectors translating the grid to itself.*

Let us see what this theorem states. Let w be an even vector (i.e., values with 0 sum). It is easy to see that only these vectors translate the grid to itself (details about isomeric transformation of the grid can be seen in, e.g., [13]). Then the distance of the points p and q is the same as the distance of their translated images p' and q':

$$d(p, q; \alpha, \beta, \gamma) = d(p', q'; \alpha, \beta, \gamma), \text{ where}$$

$p' = p + w, q' = q + w$, with $p + w = (p(1) + w(1), p(2) + w(2), p(3) + w(3))$ and similarly for $q + w$.

Since the translated lanes can also be defined and the structure of the grid is the same at the translated place, the same steps give a similar path between the translated points.

6   Conclusions

Digital distances are frequently used in image processing. The usage of non-traditional grids can have several advantages based on their better symmetric properties (e.g., they have more symmetry axes) than the traditional (square, cubic) grids. Based on this fact, the digital distances defined on them may have smaller rotational dependencies. Now the triangular grid is considered. The three types of basic neighborhood relations give more flexibility in this grid than we have on the square and hexagonal grids. The theory of neighborhood sequences were already described (see, e.g., [14]) but they could have some non-pleasant properties, for instance some of them are define not metrical distances. In this paper weighted distances are investigated. Theoretical background, such as computation of a shortest weighted path and weighted-distances are detailed. Since all of these distances are metric we recommend to use them in various applications where digital metrics are required. The next steps, to find optimal weights for various aims and to develop applications are remained for future work.

References

1. Das, P.P., Chakrabarti, P.P., Chatterji, B.N.: Generalised distances in digital geometry. Inform. Sci. 42, 51–67 (1987)
2. Das, P.P., Chakrabarti, P.P., Chatterji, B.N.: Distance functions in digital geometry. Inform. Sci. 42, 113–136 (1987)
3. Deutsch, E.S.: Thinning algorithms on rectangular, hexagonal and triangular arrays. Comm. of the ACM 15, 827–837 (1972)
4. Fouard, C., Strand, R., Borgefors, G.: Weighted distance transforms generalized to modules and their computation on point lattices. Pattern Recognition 40, 2453–2474 (2007)
5. Her, I.: Geometric transformations on the hexagonal grid. IEEE Trans. on Image Proc. 4, 1213–1221 (1995)
6. Klette, R., Rosenfeld, A.: Digital geometry. Geometric methods for digital picture analysis. Morgan Kaufmann Publishers, Elsevier Science B.V., San Francisco, Amsterdam (2004)
7. Luczak, E., Rosenfeld, A.: Distance on a hexagonal grid. Trans. on Computers C-25(5), 532–533 (1976)
8. Middleton, L., Sivaswamy, J.: Hexagonal Image Processing: A Practical Approach. Springer, London (2005)
9. Nagy, B.: Shortest path in triangular grids with neighbourhood sequences. Journal of Computing and Information Technology 11, 111–122 (2003)
10. Nagy, B.: Digital geometry of various grids based on neighbourhood structures. In: KEPAF 2007 – 6th Conference of Hungarian Association for Image Processing and Pattern Recognition, Debrecen, Hungary, pp. 46–53 (2007)
11. Nagy, B.: Distances with neighbourhood sequences in cubic and triangular grids. Pattern Recognition Letters 28, 99–109 (2007)
12. Nagy, B.: Distance with generalised neighbourhood sequences in nD and ∞D. Disc. Appl. Math. 156, 2344–2351 (2008)
13. Nagy, B.: Isometric transformations of the dual of the hexagonal lattice. In: ISPA 2009 – 6th International Symposium on Image and Signal Processing and Analysis, Salzburg, Austria, pp. 432–437 (2009)

14. Nagy, B.: Distances based on neighborhood sequences in the triangular grid. In: Computational Mathematics: Theory, Methods and Applications, pp. 313–351. Nova Science Publishers (2011)
15. Nagy, B., Strand, R., Normand, N.: A weight sequence distance function. In: Hendriks, C.L.L., Borgefors, G., Strand, R. (eds.) ISMM 2013. LNCS, vol. 7883, pp. 292–301. Springer, Heidelberg (2013)
16. Rosenfeld, A., Pfaltz, J.L.: Distance functions on digital pictures. Pattern Recognition 1, 33–61 (1968)
17. Russel, S.J., Norvig, P.: Artificial intelliegence: a modern approach. Prentice Hall (1995)
18. Sintorn, I.-M., Borgefors, G.: Weighted distance transforms in rectangular grids. In: ICIAP 2001, pp. 322–326 (2001)
19. Svensson, S., Borgefors, G.: Distance transforms in 3D using four different weights. Pattern Recognition Letters 23, 1407–1418 (2002)
20. Yamashita, M., Ibaraki, T.: Distances defined by neighborhood sequences. Pattern Recognition 19, 237–246 (1986)

An Efficient Algorithm for the Generation of Z-Convex Polyominoes[*]

Giusi Castiglione[1] and Paolo Massazza[2]

[1] Università degli Studi di Palermo, Dipartimento di Matematica ed Informatica,
Via Archirafi 34, 90123 Palermo, Italy
giusi@math.unipa.it
[2] Università degli Studi dell'Insubria, Dipartimento di Scienze
Teoriche e Applicate - Sezione Informatica, Via Mazzini 5, 21100 Varese, Italy
paolo.massazza@uninsubria.it

Abstract. We present a characterization of Z-convex polyominoes in terms of pairs of suitable integer vectors. This lets us design an algorithm which generates all Z-convex polyominoes of size n in constant amortized time.

Keywords: Polyomino, Convex Polyomino, Z-convex polyomino, Tiling, Complexity.

1 Introduction

A polyomino P [17] is a finite connected set of edge-to-edge adjacent square unit cells in the Cartesian two-dimensional plane, defined up to translations. We call size of the polyomino the number of cells that compose it. Several classes of polyominoes have been considered and extensively studied in enumerative or bijective combinatorics [7,15,3,1]. They have been studied in two-dimensional language theory where tiling systems for various classes of polyominoes have been provided [10,9]. In [2,19] the problem of reconstructing polyominoes from partial information has been addressed. Polyominoes have also received a particular attention in tiling theory see [5,25,8,23], for example.

Due to the difficulty of solving problems for the whole class of polyominoes, many subclasses have been studied, in particular by introducing some convexity constraints on the cells. The most studied polyominoes are convex polyominoes, i.e., polyominoes whose rows and columns are connected sets of cells (there are no holes between the lower cell and the upper cell of a column, and also between the leftmost cell and the rightmost cell of a row, respectively). They have been enumerated with respect to the semi-perimeter (one half the length of the boundary) but only asymptotic results are known with respect to the size [6,18]. In [14] the authors observed that convex polyominoes have the property

[*] Partially supported by Project M.I.U.R. PRIN 2010-2011: Automi e linguaggi formali: aspetti matematici e applicativi.

R.P. Barneva, V.E. Brimkov, and J. Šlapal (Eds.): IWCIA 2014, LNCS 8466, pp. 51–61, 2014.
© Springer International Publishing Switzerland 2014

that every pair of cells is connected by a monotone path (made of steps in only two directions: North and East, North and West, South and East or South and West) inside the polyomino itself. In this way, one can associate with each convex polyomino P the minimal number k of changes of direction which are necessary to ensure the existence of a monotone path between any two cells of P.

More precisely, a convex polyomino is called k-*convex* if, for every pair of its cells, there is at least one monotone path with at most k changes of direction that connects them. Very recently, an asymptotic estimate of a lower bound for the number of k-convex polyominoes with perimeter p, for any k, has been given [24]. Such a hierarchical approach provides a way to handle the general class of convex polyominoes. When the value of k is 1 we have the so-called L-convex polyominoes, where this terminology is motivated by the L-shape of the paths that connect any two cells which are not in the same column or in the same row. L-convex polyominoes have been studied from a combinatorial point of view: in [11,13] the authors enumerate them with respect to the semiperimeter and size; furthermore in [12] a reconstruction algorithm from partial information is given.

Here we are interested in the second level of such a hierarchy, that is, the class of *Z-convex polyominoes* (or, equivalently, 2-*convex* polyominoes). They have been enumerated with respect to semiperimeter in [16] and studied in [26] from the tomographic point of view, but many problems are still open. Our main motivation is to give a methodology and a characterization that allow to handle Z-convex polyominoes and at the same time represent an idea that could be generalized to k-convex polyominoes, for each k. In fact, we show a characterization of Z-convex polyominoes of size n in terms of pairs of suitable integer vectors, and we see that such a characterization directly leads to a CAT (constant amortized time) algorithm for the exhaustive generation of Z-convex polyominoes of size n. We recall that CAT algorithms for the exhaustive generation (with respect to the size) of parallelogram polyominoes [20], L-convex polyominoes [21] and for the whole class of convex polyominoes [22], have been recently proposed.

2   Preliminaries

A polyomino P is called *convex* if the intersection of P with an infinite row or column of connected cells is always connected. We indicate by $\mathrm{CPol}(n)$ the set of convex polyominoes of size n. The *size* and the *width* of P are the number $s(P)$ of its cells and the number $w(P)$ of its columns, respectively. A *path* in P is a sequence $c_1, c_2, \ldots, c_k$ of cells of P such that for all i, with $1 \leq i < k$, c_i shares one edge with c_{i+1}.

Three classical families of convex polyominoes are Ferrer diagrams (see Fig. 1(a)), stack polyominoes (see Fig. 1(b)) and parallelogram polyominoes (see Fig. 1(c)). These particular convex polyominoes are characterized with respect to the number of vertices of the minimal bounding rectangle of the polyomino which belong to the polyomino itself: three for Ferrer diagrams, two (adjacent) for stacks polyominoes and two (opposite) for parallelogram polyominoes. A polyomino is *h-centered* [4] if it contains at least one row touching both the left and the right side of the minimal bounding rectangle (see Fig. 1(d)).

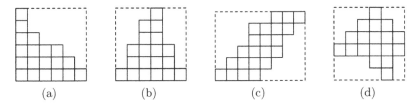

Fig. 1. (a) A Ferrer diagram; (b) a stack polyomino; (c) a parallelogram polyomino and (d) an h-centered convex polyomino

We denote by (i, j) a cell in column i and row j. Without loss of generality, we suppose that for any polyomino P one has $\min\{i|\exists j \ (i, j) \in P\} = 1$ and $\min\{j|(1, j) \in P\} = 0$.

The *vertical projection* of P is the integer vector $\pi(P) = [p_1, \ldots, p_l]$ where $l = w(P)$ and for all i, with $1 \leq i \leq l$, p_i is the number of cells of column i, $p_i = \sharp\{j|(i, j) \in P\}$. Moreover, the *position vector* of P is defined as the integer vector $\sigma(P) = [s_1, \ldots, s_l]$ where for all i, with $1 \leq i \leq l$, s_i is the y-coordinate of the bottom cell of column i, $s_i = \min\{j|(i, j) \in P\}$. Given a class of polyominoes $A \subseteq \mathrm{CPol}(n)$, notice that any $P \in A$ is univocally described by the pair of integer vectors (p, s) where $p = \pi(P)$ and $s = \sigma(P)$.

The following notations are useful when dealing with integer vectors. Let $t = [t_1, \ldots, t_l]$. We denote by $x^{[k]}$ the vector $[\underbrace{x, \ldots, x}_{k}]$, by $\cdot$ the *catenation product* of vectors and we let $t_{<i} = [t_1, \ldots, t_{i-1}]$ $(t_{\leq i} = [t_1, \ldots, t_i])$ and $t_{>i} = [t_{i+1}, \ldots, t_l]$ $(t_{\geq i} = [t_i, \ldots, t_l])$.

We say that t is *unimodal* if it can be written as $t = t_{\leq i} \cdot t_{>i}$, $1 \leq i \leq l$, with $t_j \leq t_{j+1}$ for each $1 \leq j < i$, $t_i > t_{i+1}$ and $t_j \geq t_{j+1}$ for each $i < j < l$. We say that t is *concave* if it can be written as $t = t_{\leq i} \cdot t_{>i}$, $1 \leq i \leq l$, with $t_j \geq t_{j+1}$ for each $1 \leq j < i$, $t_i < t_{i+1}$ and $t_j \leq t_{j+1}$ for each $i < j < l$.

If P is a convex polyomino one can easily observe that the vector of y-coordinates of the top cells of the columns of P is concave and the vector of y-coordinates of the bottom cells of the columns is unimodal, or vice versa. Note that a constant vector (for example, when P is a stack polyomino) can be, equivalently, considered both unimodal and concave.

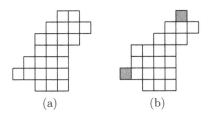

Fig. 2. (a) A Z-convex polyomino, (b) a 3-convex polyomino

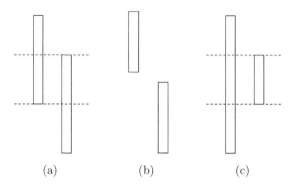

(a)   (b)   (c)

Fig. 3. (a) Columns with a horizontal intersection, (b) horizontally disjoint columns, (c) column horizontally included

3   Z-Convex Polyominoes

A convex polyomino P is said *Z-convex* if any two cells of P are connected by a path in P with at most two changes of direction. Figure 2 shows a Z-convex polyomino together with a polyomino which is 3-convex but not Z-convex (every path between the highlighted cells has at least three changes of direction).

We denote by $ZPol(n)$ the set of Z-convex polyominoes of size n. In this section we give a characterization of Z-convex polyominoes in terms of position vectors and vertical projections and we give some basic properties which are fundamental for the generation of $ZPol(n)$. Firstly, we give a basic definition.

Definition 1. *Let P be a convex polyomino with $\sigma(P) = [s_1, \ldots, s_l]$ and $\pi(P) = [p_1, \ldots, p_l]$. Let $1 \leq i, j \leq l$, we say that*

- *columns i and j of P have a horizontal intersection (see Fig. 3(a)) iff $s_j < s_i \leq s_j + p_j - 1 < s_i + p_i - 1$ or $s_i < s_j \leq s_i + p_i - 1 < s_j + p_j - 1$;*
- *columns i and j of P are horizontally disjoint (see Fig. 3(b)) iff $s_i > s_j + p_j - 1$ or $s_j > s_i + p_i - 1$;*
- *column i horizontally includes column j of P (see Fig. 3(c)) iff $s_i \leq s_j$ and $s_i + p_i - 1 \geq s_j + p_j - 1$.*

The previous definition can be used to characterize Z-convex polyominoes, as stated in the following theorem.

Theorem 1. *Let P be a convex polyomino. Then P is Z-convex if and only if for all i, j, with $1 \leq i < j \leq l$, if columns i and j are horizontally disjoint then there exists k, with $i < k < j$, such that column k horizontally includes both columns i and j.*

Proof. Le P be a Z-convex polyomino and i and j two horizontally disjoint columns of P. Let us consider the top cell $x = (i, s_i + p_i - 1)$ of column i and the bottom cell $y = (j, s_j)$ of column j. If $s_i > s_j + p_j - 1$ (see Fig. 3(b)) the only way to connect x to y by a path with at most two changes of direction is a path that starts with a horizontal right step and changes direction in cells $(k, s_i + p_i - 1)$ and (k, s_j), for a certain k. Then there exists in P a column k with $s_k \leq s_j$ and $p_k \geq p_j + p_i$, and so column k includes both columns i and j. Vice versa, for any pair of cells $x = (i, h)$ and $y = (j, g)$ if columns i and j have a horizontal intersection with $s_i \leq s_j + p_j - 1$ (see Fig. 3(a)) then there exists a row in P, formed by the cells $(i, m), (i + 1, m), \ldots, (j, m)$, with $h \leq m \leq g$, that crosses both columns i and j. Then, there is a path which connects x to y and has exactly two changes of direction, in cells (i, m) and (j, m). Otherwise, if i and j are horizontally disjoint, by hypothesis we can consider the path that starts with an horizontal right step and changes direction in cells (k, h) and (k, g). $\square$

Let $\sigma(P) = [s_1, \ldots, s_l]$ and define $\mathrm{TOP}(P) = [s_1 + p_1 - 1, \ldots, s_l + p_l - 1]$. As observed in the previous section, if P is a convex polyomino then $\sigma(P)$ is concave and $\mathrm{TOP}(P)$ unimodal, or vice versa. Without loss of generality, from here on we consider convex polyominoes of the first kind, i.e. $\sigma(P)$ is concave and $\mathrm{TOP}(P)$ is unimodal; we call *descending* this kind of polyominoes.

Thus, let h be the first index such that $s_{h-1} < s_h$ or column h is horizontally disjoint from one of the first $h - 1$ columns and let k be the smallest index such that column k is not included in column $k + 1$. Trivially, one has $k < h$.

We write $\sigma(P) = s_{\leq h} \cdot s_{>h}$ and $\mathrm{TOP}(P) = t_{\leq k} \cdot t_{>k}$. It follows a decomposition of a Z-convex polyomino as stated in the following proposition.

Proposition 1. *Let P be a Z-convex polyomino with $\sigma(P) = s_{\leq h} \cdot s_{>h}$ and $\mathrm{TOP}(P) = t_{\leq k} \cdot t_{>k}$. Then P can be decomposed into three (possibly null) parts: a left stack polyomino L (the first k columns) and a right stack polyomino R (the last $w(P) - h + 1$ columns) joined by a central parallelogram polyomino C (from column $k + 1$ to column $h - 1$) with the following properties:*

- *C is an h-centered convex polyomino (see Fig. 4(a)).*
- *L and C form an h-centered polyomino (see Fig. 4(b)).*
- *C and R form an h-centered polyomino (see Fig. 4(c)).*
- *if there exist in P two horizontally disjoint columns then they belong to L and to R, respectively (see the highlighted columns in Fig. 4(d)).*

Proof. Consider the fourth property. Let i, j be two horizontally disjoint columns, with $i < j$, and suppose that $s_i > s_j + p_j - 1$. By Theorem 1 there exists a column k, with $i < k < j$, that includes both columns i and j. Moreover, for convexity reasons the y-coordinate of the top cell of a column h to the right of column j must satisfy the relation $\mathrm{TOP}[h] \leq \mathrm{TOP}[j]$. So, columns h and i are horizontally disjoint and then there is g, with $i < g < j < h$, such that column g includes both columns i and h. If one had $s_h < s_j$ then it would be impossible to have $s_g \leq s_h$ due to the convexity constraint. Therefore, any column h to the right of column j is horizontally included in column j, and column j belongs to the right stack R. In the same way one can prove that column i belongs to the left stack L. The others properties are proved as a consequence. $\square$

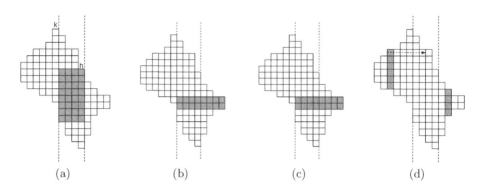

Fig. 4. Properties of the decomposition of a Z-convex polyomino

Let P be a Z-convex polyomino with decomposition $P = L \cdot C \cdot R$ provided by Proposition 1. Note that the left stack L is never null and consists of the first k columns of P, $k \geq 1$. Then we can give the following definition.

Definition 2. *Let* $s = \sigma(P) = [s_1, \ldots, s_l]$ *and* $s_{\leq k} = 0^{[j_1]} \cdot x_2^{[j_2]} \cdots x_s^{[j_s]}$, *with* $0 > x_2 \cdots > x_s$. *We define the integer vector* BOTTOMCELL *as* BOTTOMCELL$[e] = \sum_{i \leq e} j_i$, *with* $1 \leq e \leq s$.

In other words, if $B = \{0, x_2, \ldots, x_s\}$ then the eth entry of BOTTOMCELL is the index of the rightmost column of the left stack L such that the y-coordinate of its bottom cell is the eth value in B.

Furthermore if $C \neq \emptyset$ we can give the following definition.

Definition 3. *Let* m *be an integer, with* $k < m < h$, *such that the first* m *columns of* P *form an* h-*centered polyomino with horizontal intersection given by a rectangle of height* p_1 *(the height of the first column of* L*). We set* DOM$[m] = e$, *if column* m *horizontally includes column* BOTTOMCELL$[e]$ *but does not horizontally include column* BOTTOMCELL$[e+1]$. *We define also the integer vector* ISDOM *such that* ISDOM$[e] = m$, *with* $1 \leq e \leq s$, *if column* DOM$[m] = e$ *In such a case we say that column* m *dominates column* BOTTOMCELL$[e]$.

Example 1 illustrates the vectors BOTTOMCELL and ISDOM associated with the polyomino of Fig. 4. The importance of dominating columns is pointed out by the following lemma.

Lemma 1. *If two columns* i, j *of a Z-convex polyomino, with* $i \leq k$ *and* $j > h$, *are horizontally disjoint then column* j *is horizontally included in the column that dominates column* i.

Proof. Recall that if columns i and j are horizontally disjoint then there exists a column that includes them. The column m which dominates column i is the rightmost that includes i and so it is the column with the lowest bottom cell among the columns which include column i. Thus m includes also column j. □

Example 1. Let us consider the Z-convex polyomino P in Fig. 4(d). We have

$$\sigma(P) = [0, -1, -1, -2, -2, -3, -6, -8, -9, -10, -8, -5, -4, -4]$$

and

$$\text{TOP}(P) = [2, 3, 4, 4, 5, 7, 6, 4, 4, 1, -1, -2, -3, -3].$$

Furthermore, the left stack has width $k = 6$, with $\text{BOTTOMCELL} = [1, 3, 5, 6]$. Moreover, one has $\text{ISDOM} = [9, 9, 7, 6]$.

Let us observe that column 12 and column $\text{BOTTOMCELL}[2]$ (highlighted in Fig. 4(d)) are horizontally disjoint and that column $\text{ISDOM}[2]$ (pointed by the arrow) dominates column $\text{BOTTOMCELL}[2]$, i.e. column 9 dominates column 3 and horizontally includes column 12.

4   The Exhaustive Generation of ZPol(n)

In this section we give an outline of an algorithm which works column by column and generates all Z-convex polyominoes of size n. The algorithm is based on an inductive approach: at step i it assumes that a Z-convex polyomino P_{i-1} with $i-1$ columns, vertical projection $[p_1, \ldots, p_{i-1}]$, position vector $[s_1, \ldots, s_{i-1}]$ and size $n-r$ has been generated and it determines all and only those ith columns (of size at most r) that can extend it. More precisely, it computes all the integer pairs (a, b) such that $[p_1, \ldots, p_{i-1}, a]$ and $[s_1, \ldots, s_{i-1}, b]$ denote a Z-convex polyomino. Obviously, by convexity reasons, the values a, b must satisfy the relations

$$b \leq s_{i-1} + p_{i-1} - 1, \quad b + a - 1 \geq s_{i-1} .$$

Let us call $\text{H-CENTERED}(i, r, d)$ a recursive procedure that, given a Z-convex polyomino P_{i-1} with size $n - r$ and horizontal intersection of height d, generates all Z-convex polyominoes with prefix P_{i-1} by computing all the possible ith columns (identified by the size a and the position b) and making a recursive call for each of them. The crucial points of the algorithm are described below, where all the cases which may arise when adding column i, of size a and position b, are considered (recall the decomposition $P = L \cdot C \cdot R$).

1. Column i horizontally contains all the previous columns. In this case P_i is a prefix of the left stack L and $d = p_1$. So, the procedure updates the vector BOTTOMCELL and makes a recursive call $\text{H-CENTERED}(i + 1, r - a, p_1)$;
2. P_{i-1} is a prefix of L, P_i is h-centered ($b + a > 0$) and column i does not horizontally includes column $i-1$. In this case column i is the first column of C and the vectors DOM and ISDOM are possibly updated before the recursive call $\text{H-CENTERED}(i + 1, r - a, \min(d, b + a))$;
3. P_{i-1} is a prefix of L and $b > s_{i-1}$ or $b = s_{i-1}$ and P_i is not h-centered ($b + a < 0$). In this case C is null and column i is the first column of R. We then proceed by calling a simple procedure $\text{R}(i + 1, r - a, a)$ to generate all right stacks of height at most a and size $r - a$;

4. P_{i-1} is not a prefix of L, is h-centered $(d > 0)$ and $b + a > 0$. In this case column i belongs either to C (if $b \leq s_{i-1}$ – the vectors DOM and IS-DOM are possibly updated) or to R (if $b > s_{i-1}$), and the recursive call H-CENTERED$(i + 1, r - a, \min(d, b + a))$ occurs;

5. P_{i-1} is not a prefix of L, is h-centered $(d > 0)$ and $b + a \leq 0$. In this case column i is the first column of R (there must be a column $j < i$ including i and any column h horizontally disjoint from i), the vectors DOM and ISDOM are not updated and the recursive call H-CENTERED$(i + 1, r - a, 0)$ occurs.

6. P_{i-1} is not h-centered $(d = 0)$, that is, at least one column of R has been already generated. In this case we have to generate a suffix of R. The position b of column i has to be determined by exploiting the vectors ISDOM and DOM and the recursive call H-CENTERED$(i + 1, r - a, 0)$ occurs.

The previous description provides us with the high level behaviour of the algorithm. Nevertheless, a detailed analysis of the steps associated with Case 2 (the generation of the first column of C) is needed to point out the special role played by the vectors ISDOM and DOM, in particular how they can be updated in time $O(1)$. So, let us see how H-CENTERED$(k + 1, r, p_1)$ works when P_k is a (left) stack.

First, we point out that procedure H-CENTERED examines the integer pairs (column-size, bottomcell-position) according to the following order:

$$(a, b) < (a', b') \quad \text{iff} \quad a < a' \vee a = a' \wedge b > b'.$$

This corresponds to two nested loops, the outer associated with the size a (increasing form 1 to r) and the inner associated with the position b (decreasing from $p_k + s_k - a$ to $s_k - a + 1$).

So, the first column that can belong to C corresponds to the pair $(-s_k + 1, s_k)$ (a column associated with a pair $(a', b') < (-s_k + 1, s_k)$ is necessarily the first column of R). Such a column may possibly dominate only column BOTTOMCELL[1]: this happens if $p_j = 1$ with $j = $ BOTTOMCELL[1], that is, if $1^{[j]}$ is a prefix of $p_{\leq k}$. In this case we set DOM$[k + 1] = 1$ and ISDOM$[1] = k + 1$. More generally, an inductive approach is used to determine which column is dominated by the $(k + 1)$-th column associated with (a, b): it is sufficient to remember the entries $e' = $ DOM$[k+1]$ and $e'' = $ DOM$[k+1]$ associated with the pair (a', b') which immediately precedes (a, b) and with the pair $(a-1, p_k + s_k - a + 1)$ (the column of size $a - 1$ in the highest position), respectively. Indeed, if $b + a - 1 < s_k + b_k$ a simple check let us to see whether to set either DOM$[k+1] = e'$ or DOM$[k+1] = e' + 1$ or DOM$[k+1] = e' - 1$, whereas if $b + a - 1 = s_k + b_k$ one has either DOM$[k+1] = e''$ or DOM$[k + 1] = e'' + 1$ (the vector ISDOM is updated accordingly).

By the same method we can operate in Case 4, when all the subsequent columns j which belong to the parallelogram C, with $j > k + 1$, are placed at a position for which P_j is h-centered.

Lastly, we illustrate how the call H-CENTERED(i, r, d) works in Case 5. First, since $b + a \leq 0$, the size a of the first column of R is at most $-s_{i-1}$ (the pairs (a, b) are still generated according to the order $<$ previously defined). By Theorem 1, the highest position for a column i of size a and such that P_i is not h-centered (but still Z-Convex) is the value b which satisfies $b + a = 0$, provided that there is $j < i$ such that column j horizontally includes both column BOTTOMCELLl[1] and column i. This can be checked by accessing column j, where $j = $ ISDOM[1] and testing whether $s_j \leq -a$. If $s_1 > -a$ then the position $-a + 1$ (analysed in the previous iteration over b) is the lowest admissible for a column of size a. Thus, $b = -a$ can be considered as the basis of our inductive approach for generating valid positions. More generally, in order to check whether $-a - q$ can be a valid position for a column of size a we need only to remember the index e' such that at the previous iteration (the position $-a - q + 1$) column BOTTOMCELL[e'] was the rightmost column horizontally disjoint from column i. If column BOTTOMCELL[$e' + 1$] is now the rightmost column horizontally disjoint from column i then we read ISDOM[$e' + 1$] to get the index j and make the test $s_j \leq -a - q$. Otherwise the index $j = $ BOTTOMCELL[e'] is used also for testing the current position.

We point out that procedure H-CENTERED works similarly also in Case 6.

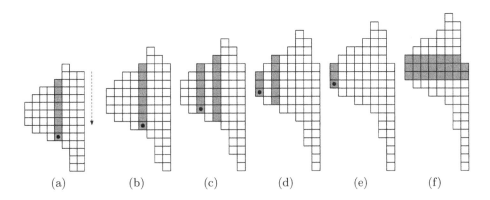

Fig. 5. Some iterations of H-CENTERED$(8, 13, 3)$

Example 2. In Fig. 5 we show the first 6 iterations for b when $a = 13$ of the call H-CENTERED$(8, 13, 3)$. In (a), since $p_8 = p_7$ and $s_8 = s_7$ one has DOM[8] = DOM[7] = e = 3 and the current value (7) of ISDOM[3] is replaced by 8. Nothing changes at the 2nd iteration (b). In (c) since TOP[8] < TOP[BOTTOMCELL[e]] then $e = e - 1 = 3 - 1 = 2$, DOM[8] = 2 and ISDOM[2] = 8 and the old value ISDOM[3] = 7 is restored. In (d) since TOP[8] < TOP[BOTTOMCELL[e]] then $e = e - 1 = 2 - 1 = 1$, DOM[8] = 1 and ISDOM[1] = 8, Nothing changes at the 5th iteration (e). At the 6th iteration (f) column 8 reduces the horizontal intersection then it does not dominate any column of the left stack L.

5   Complexity

We provide in this section some remarks that let us understand why the proposed algorithm is CAT. The algorithm has a recursive structure. Its execution can be described by a tree where each node at level i corresponds to a call of a procedure which works on column $i + 1$ (the root is associated with the call H-CENTERED$(1, n, n)$). Moreover, the outdegree of each internal node is at least 2. This implies that the number of internal nodes is $O(N)$ where N is the number of leaves, that is, the numbers of polyominoes in ZPol(n).

Hence, it follows that the algorithm is CAT if we show that each call (working on a given column, say i) has a running time $O(K)$, where

$$K = \#\{(a, b)|P_i \text{ identified by } [p_1, \ldots, p_{i-1}, a], [s_1, \ldots, s_{i-1}, b] \text{ is Z-convex}\}.$$

Indeed, the strategy that procedure H-CENTERED adopts for determining all the admissible pairs (a, b) (representing the size and the position of the column to be generated, respectively) guarantees that.

As shown in the previous section the structure of the procedure consists of two nested cycles, the outer one associated with the size a of the column, the inner one associated with the position b of the bottom cell. For each size a (analysed in increasing order) the admissible positions are generated downwards, with a constant number of operations associated with each value. As soon as a value b' is reached such that the two integer vectors $[p_1, \ldots, p_{i-1}, a]$ and $[s_1, \ldots, s_{i-1}, b']$ do not denote a Z-convex polomino, a break in the inner loop occurs and a new iteration of the outer loop starts (for the size $a + 1$). The key observation is that if b' is not a valid position for the current size then, for any integer $q > 0$, also $b' - q$ is not valid. Note also that if we can not find a valid position for a column of size a, then a valid position for a column of size $a + j$, for all $j > 0$, does not exist. So Procedure H-CENTERED has to return the control to the caller as soon as such a size is reached.

6   Conclusions

In this paper we have shown that, by representing Z-convex polyominoes as pairs of suitable integer vectors, we can easily obtain a CAT algorithm for the exhaustive generation of ZPol(n). It is also straightforward to see that our approach can be exploited for solving the membership problem for ZPol(n). More precisely, having as input a convex polyomino $P \in$ CPol(n) individuated by $\pi(P)$ and $\sigma(P)$, we test whether $P \in$ ZPol(n) in time $O(l)$ where $l = w(P)$. It is also natural to ask whether such an approach can be used to test k-convexity for a generic k, and more generally to design a CAT generation algorithm for k-convex polyominoes. So, a first step in this direction is to find a characterization of k-convex polyominoes in terms of vertical projections and position vectors.

References

1. Barcucci, E., Frosini, A., Rinaldi, S.: On directed-convex polyominoes in a rectangle. Discrete Mathematics 298(1-3), 62–78 (2005)
2. Barcucci, E., Lungo, A.D., Nivat, M., Pinzani, R.: Reconstructing Convex Polyominoes from Horizontal and Vertical Projections. Theor. Comput. Sci. 155(2), 321–347 (1996)

3. Barcucci, E., Lungo, A.D., Pergola, E., Pinzani, R.: ECO: a methodology for the Enumeration of Combinatorial Objects. J. of Diff. Eq. and App. 5, 435–490 (1999)
4. Battaglino, D., Fedou, J.M., Frosini, A., Rinaldi, S.: Encoding Centered Polyominoes by Means of a Regular Language. In: Mauri, G., Leporati, A. (eds.) DLT 2011. LNCS, vol. 6795, pp. 464–465. Springer, Heidelberg (2011)
5. Beauquier, D., Nivat, M.: On Translating One Polyomino to Tile the Plane. Discrete & Computational Geometry 6, 575–592 (1991)
6. Bender, E.A.: Convex n-ominoes. Discrete Math. 8, 219–226 (1974)
7. Bousquet-Mélou, M.: A method for the enumeration of various classes of column-convex polygons. Discrete Math. 154(1-3), 1–25 (1996)
8. Brlek, S., Provençal, X., Fedou, J.-M.: On the tiling by translation problem. Discrete Applied Mathematics 157(3), 464–475 (2009)
9. Brocchi, S., Frosini, A., Pinzani, R., Rinaldi, S.: A tiling system for the class of L-convex polyominoes. Theor. Comput. Sci. 475, 73–81 (2013)
10. Carli, F.D., Frosini, A., Rinaldi, S., Vuillon, L.: On the Tiling System Recognizability of Various Classes of Convex Polyominoes. Ann. Comb. 13, 169–191 (2009)
11. Castiglione, G., Frosini, A., Munarini, E., Restivo, A., Rinaldi, S.: Combinatorial aspects of L-convex polyominoes. Eur. J. Comb. 28(6), 1724–1741 (2007)
12. Castiglione, G., Frosini, A., Restivo, A., Rinaldi, S.: A Tomographical Characterization of L-Convex Polyominoes. In: Andrès, É., Damiand, G., Lienhardt, P. (eds.) DGCI 2005. LNCS, vol. 3429, pp. 115–125. Springer, Heidelberg (2005)
13. Castiglione, G., Frosini, A., Restivo, A., Rinaldi, S.: Enumeration of L-convex polyominoes by rows and columns. Theor. Comput. Sci. 347(1-2), 336–352 (2005)
14. Castiglione, G., Restivo, A.: Reconstruction of L-convex Polyominoes. Electronic Notes in Discrete Mathematics 12, 290–301 (2003)
15. Delest, M.P., Viennot, G.: Algebraic Languages and Polyominoes Enumeration. Theor. Comput. Sci. 34, 169–206 (1984)
16. Duchi, E., Rinaldi, S., Schaeffer, G.: The number of Z-convex polyominoes. Advances in Applied Mathematics 40(1), 54–72 (2008)
17. Golomb, W.S.: Checker Boards and Polyominoes. The American Mathematical Monthly 61, 675–682 (1954)
18. Klarner, D.A., Rivest, R.R.: Asymptotic bounds for the number of convex n-ominoes. Discrete Math. 8, 31–40 (1974)
19. Kuba, A., Balogh, E.: Reconstruction of convex $2D$ discrete sets in polynomial time. Theor. Comput. Sci. 283(1), 223–242 (2002)
20. Mantaci, R., Massazza, P.: From Linear Partitions to Parallelogram Polyominoes. In: Mauri, G., Leporati, A. (eds.) DLT 2011. LNCS, vol. 6795, pp. 350–361. Springer, Heidelberg (2011)
21. Massazza, P.: On the generation of L-convex polyominoes. In: Proc. of GASCom 2012, Bordeaux, June 25-27 (2012)
22. Massazza, P.: On the Generation of Convex Polyominoes. Discrete Applied Mathematics (to appear)
23. Massé, A.B., Garon, A., Labbé, S.: Combinatorial properties of double square tiles. Theor. Comput. Sci. 502, 98–117 (2013)
24. Micheli, A., Rossin, D.: Counting k-Convex Polyominoes. Electr. J. Comb. 20(2) (2013)
25. Ollinger, N.: Tiling the Plane with a Fixed Number of Polyominoes. In: Dediu, A.H., Ionescu, A.M., Martín-Vide, C. (eds.) LATA 2009. LNCS, vol. 5457, pp. 638–647. Springer, Heidelberg (2009)
26. Tawbe, K., Vuillon, L.: 2L-convex polyominoes: Geometrical aspects. Contributions to Discrete Mathematics 6(1) (2011)

Using a Topological Descriptor
to Investigate Structures of Virus Particles

Lucas M. Oliveira[1], Gabor T. Herman[1], Tat Yung Kong[1,2],
Paul Gottlieb[3], and Al Katz[4]

[1] Computer Science PhD Program, The Graduate Center, City University of New York,
365 Fifth Avenue, New York, NY 10016, U.S.A.
lmoliveira@gmail.com, gabortherman@yahoo.com
[2] Computer Science Department, Queens College,
City University of New York, 65-30 Kissena Boulevard, Flushing, NY 11367, U.S.A.
ykong@cs.qc.cuny.edu
[3] Department of Pathobiology of the Sophie Davis School of Biomedical Education,
The City College of New York, 138th Street and Convent Avenue, New York, NY 10031, U.S.A.
pgottl@med.cuny.edu
[4] Department of Physics, The City College of New York, 160 Convent Avenue,
New York, NY 10031, U.S.A.
akatz@sci.ccny.cuny.edu

Abstract. An understanding of the three-dimensional structure of a biological macromolecular complex is essential to fully understand its function. A component tree is a topological and geometric image descriptor that captures information regarding the structure of an image based on the connected components determined by different grayness thresholds. We believe interactive visual exploration of component trees of (the density maps of) macromolecular complexes can yield much information about their structure. To illustrate how component trees can convey important structural information, we consider component trees of four recombinant procapsids of a bacteriophage (cystovirus $\phi6$), and show how differences between the component trees reflect the fact that each non-wild-type mutant of the procapsid has an incomplete set of constituent proteins.

Keywords: Topological descriptor, Macromolecule, Rooted tree, Segmentation, Digital image.

1 Introduction

The three-dimensional structure of a biological macromolecular complex is tightly related to its function within a cell, and knowledge of the structure is often necessary to understand important processes such as viral RNA packing. One of the problems of structural biology is that of deducing details of the structure of a macromolecular complex from its density map (which is a three-dimensional image of the complex obtained, e.g., by reconstruction from electron micrographs). Density maps of many macromolecular complexes are now available from public repositories such as EMDataBank [5,12].

Density maps are usually visualized by displaying two-dimensional slices, surface-renderings, or volume-renderings. However, we believe much structural information can be discovered by interactive visual exploration of *component trees* of density maps.

R.P. Barneva, V.E. Brimkov, and J. Šlapal (Eds.): IWCIA 2014, LNCS 8466, pp. 62–75, 2014.
© Springer International Publishing Switzerland 2014

A component tree is a compact image descriptor that manifests structural relationships between different parts of the image. It captures topological and geometric information based on the connected components determined by different grayness thresholds. Over the years, component trees have been used in a variety of image processing algorithms, including algorithms that perform image segmentation [1,3,10,13,19,20,24], image simplification [21], and object identification [14].

When constructed for macromolecular density maps, appropriately simplified component trees can convey essential structural information about a specimen that is independent of the resolution of its density map and the process (e.g., cryo-electron microscopy or X-ray crystallography) used to obtain that map.

An important step in determining the structure of a macromolecular complex is to identify its constituent substructures. We believe this is an example of a problem in which information conveyed by component trees may be particularly useful. To illustrate the relevance of component trees in this context, we consider four recombinant procapsids of a bacteriophage ($\phi6$) and show how differences between simplified component trees of the four versions reflect the fact that each non-wild-type mutant has an incomplete set of constituent proteins.

2 Digital Spaces and Digital Pictures

Let V be a nonempty finite set and π a symmetric irreflexive binary relation on V. The elements of V are called *spels* (an abbreviation of "spatial elements"), and can be seen as generalizations of pixels and voxels [7]. If $(c,d) \in \pi$, then we say that c and d are π-*adjacent*. Let A be a subset of V. For any c and d in A, a sequence $\left\langle d^{(0)}, \ldots, d^{(K)} \right\rangle$ of elements of A is said to be a π-*path in A connecting c to d* if $d^{(0)} = c$, $d^{(K)} = d$, and, for $0 \leq k < K$, $d^{(k)}$ is π-adjacent to $d^{(k+1)}$. If there is a π-path in A connecting c to d, then we say that c is π-*connected in A to d*. The subset A of V is said to be π-*connected* if, for all c and d in A, c is π-connected in A to d. Note that the empty set is π-connected and the set $\{c\}$ is π-connected, for any c in V. If V is π-connected, then we call the pair (V, π) a *digital space*.

A *digital picture over the digital space* (V, π) is a triple (V, π, f), where f is a function that maps V into the real numbers [7]. For each $c \in V$ we say that $f(c)$ is the *intensity* level of c. For any real number t, the t-*superlevel set of (V, π, f)* is $V_{f,t} = \{c \in V \mid t \leq f(c)\}$ [4].

Let A and C be subsets of V. We say that C is a π-*component* of A if

1. C is a nonempty π-connected subset of A and
2. for all π-connected subsets B of A, if $C \subseteq B$, then $B = C$.

For any digital picture (V, π, f) and any $c \in V$, there is a unique π-component of $V_{f,f(c)}$ that contains c and it is given by

$$C_{(V,\pi,f)}(c) = \left\{ d \in V \mid c \text{ is } \pi\text{-connected in } V_{f,f(c)} \text{ to } d \right\}. \tag{1}$$

The *set of components of* (V, π, f) is

$$\mathfrak{C}_{(V,\pi,f)} = \left\{ C_{(V,\pi,f)}(c) \mid c \in V \right\}. \tag{2}$$

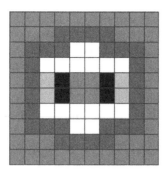

Fig. 1. A Digital Picture (V, π, f). Here V is the set of all small squares (*pixels*), π is the binary relation (on V) of having exactly one edge in common (*edge-adjacency*), and f maps elements of V to 0 (black), 1 (dark gray), 2 (gray), 3 (light gray), and 4 (white).

It is important to note that $V \in \mathfrak{C}_{(V,\pi,f)}$ and that any two members of $\mathfrak{C}_{(V,\pi,f)}$ are disjoint unless one is a subset of the other. Fig. 1 shows a digital picture (V, π, f) in which V is the set of all small squares (*pixels*), π is the binary relation (on V) of having exactly one edge in common (*edge-adjacency*), and f maps elements of V to 0, 1, 2, 3, and 4. In Fig. 1, the intensity levels 0, 1, 2, 3, and 4 are shown respectively as black, dark gray, gray, light gray, and white.

3 Component Trees

A *rooted tree* T is a pair (N, E), where N is a finite set of *nodes* and E is a set of *edges* such that the following conditions hold: (1) Each edge is an ordered pair of distinct nodes in which the first node is called the *parent* node and the second node the *child* node of the edge; (2) every node, with the exception of one node called the *root*, is a child node of just one edge; (3) if E is regarded as a binary relation on N, then the reflexive transitive closure of E is a partial order on N.

If m and n are nodes such that $m = n$ or m precedes n in the partial order, then n is called a *descendant* of m and m is called an *ancestor* of n. Readily, every node in N is a descendant of the root. We say m is a *proper descendant* (respectively, *proper ancestor*) of n if m is a descendant (respectively, an ancestor) of n and $m \neq n$.

If p and c are respectively the parent node and the child node of some edge, then we also call p the parent of the node c and call c a child of the node p. A node that has no children is called a *leaf*.

A *labeled tree* is a triple (N, E, Ω), where (N, E) is a rooted tree and Ω is a function that maps N into the real numbers.

3.1 The Component Tree of a Digital Picture

For any digital picture (V, π, f), the *component tree of* (V, π, f) is the labeled tree $T_{(V,\pi,f)} = \left(N_{(V,\pi,f)}, E_{(V,\pi,f)}, \Omega_{(V,\pi,f)} \right)$ for which

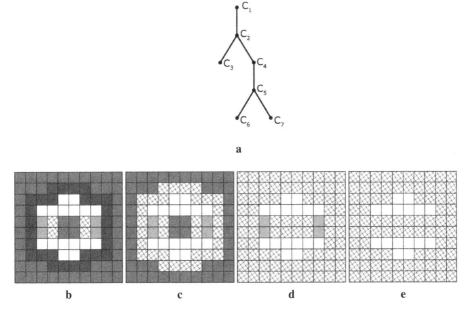

a

b   **c**   **d**   **e**

Fig. 2. The Components of the Digital Picture in Fig. 1. (a) shows the component tree. In (b), (c), (d), and (e), the cross-hatched pixels are the pixels whose intensities are below the threshold levels 1, 2, 3, and 4, respectively. Each non-cross-hatched pixel is colored dark gray, gray, light gray, or white according to whether its intensity is 1, 2, 3, or 4. Each component of the *non-cross-hatched part* of (b), (c), (d), and (e) is a non-root node of the tree (a): The components C_6 and C_7 (each of which consists of 8 pixels of intensity 4) are shown in (e); the component C_5 (which consists of the 20 pixels of intensity 3 or 4) is shown in (d); the components C_3 (44 pixels of intensity 2) and C_4 (24 pixels of intensity 2, 3, or 4) are shown in (c); the component C_2 (96 pixels of intensity ≥ 1) is shown in (b).

1. the set of nodes $N_{(V,\pi,f)}$ is $\mathfrak{C}_{(V,\pi,f)}$,
2. the root is V,
3. for any C and D in $N_{(V,\pi,f)} = \mathfrak{C}_{(V,\pi,f)}$, C is an ancestor of D if, and only if, D is a subset of C, and
4. for any $C \in N_{(V,\pi,f)}$,

$$\Omega_{(V,\pi,f)}(C) = \min\{f(e)\,|\,e \in C\}. \tag{3}$$

Recall that any two members of $\mathfrak{C}_{(V,\pi,f)}$ are disjoint unless one is a subset of the other. This implies that if $D \in \mathfrak{C}_{(V,\pi,f)}$ is a subset both of $C_1 \in \mathfrak{C}_{(V,\pi,f)}$ and of $C_2 \in \mathfrak{C}_{(V,\pi,f)}$ then one of C_1 and C_2 must be a subset of the other. So for each $D \in \mathfrak{C}_{(V,\pi,f)} - \{V\}$ there exists a unique smallest member B of $\mathfrak{C}_{(V,\pi,f)}$ that contains D as a proper subset, and it is easy to see that B is the parent of the node D in the component tree $T_{(V,\pi,f)}$. The tree as defined above is very similar to the *foreground history tree* of [22], and is isomorphic to the *join tree* of [2] if f is 1-to-1. The component tree of the digital picture of Fig. 1 is shown in Fig. 2(a), but without indicating the labels provided by Ω.

A component tree of a digital picture (V, π, f) can be created by thresholding that digital picture at every intensity level t which occurs in the picture. Thresholding at intensity level t produces the superlevel set $V_{f,t}$ each of whose π-components is a node of the component tree.

For example, if we threshold the digital picture (V, π, f) of Fig. 1 at intensity level $t = 0$ we obtain only one component, which is the root C_1 of the tree shown in Fig. 2(a); this component is equal to the set V.

When (V, π, f) is thresholded at the intensity level $t = 1$, there is again just one component, which is the node C_2 of Fig. 2(a). The component C_2 is shown in Fig. 2(b) where the four pixels of V that do not belong to C_2 (the four black pixels in Fig. 1) are cross-hatched. When the picture is thresholded at the intensity level $t = 2$, all pixels in the cross-hatched parts of Fig. 2(c) have intensity levels that are below the threshold. The remaining pixels consist of two components that can be seen in Fig. 2(c): One component, C_3, comprises the 44 dark gray pixels around the edge of the picture; the other component, C_4, comprises 24 pixels in the central part of the picture. The components obtained when the picture is thresholded at the intensity level $t = 3$ (the node C_5) and the level $t = 4$ (the nodes C_6 and C_7) can be seen in Figs. 2(d) and 2(e), respectively.

When the number of intensity levels is large the algorithm presented in [15] is a much more efficient way to construct the component tree. That algorithm does not involve thresholding the digital picture at every intensity level which occurs in the picture, but processes the picture elements in decreasing order of their intensity and uses Tarjan's union-find algorithm [23] to build the tree from the bottom up.

We have just described two ways to construct the component tree $T_{(V, \pi, f)}$ of any digital picture (V, π, f). Conversely, in any given digital space (V, π), any digital picture (V, π, f) can be recovered from its component tree $T_{(V, \pi, f)}$. To prove this, we proceed as follows.

We assume that we are given a digital space (V, π) and a labeled tree $T = (N, E, \Omega)$ that is the component tree $T_{(V, \pi, f)}$ of an unknown digital picture (V, π, f). What we need to show is how to determine, for any $d \in V$, the value of $f(d)$.

For this purpose, let $\mathbf{node}_T(d)$ denote the *smallest* node of T that contains d. We claim $\mathbf{node}_T(d) = C_{(V, \pi, f)}(d)$. To see this, note that (by the definition of the component tree $T_{(V, \pi, f)} = T$) the component $C_{(V, \pi, f)}(d)$ is a node of T that contains d, and if $C_{(V, \pi, f)}(c)$ is any node of T that contains d then (1) and the definition of $V_{f, f(c)}$ imply that $f(c) \leq f(d)$ and hence, by Prop. 2 of [9], that $C_{(V, \pi, f)}(d)$ is a subset of $C_{(V, \pi, f)}(c)$. Since $\mathbf{node}_T(d) = C_{(V, \pi, f)}(d)$, it follows from Prop. 3 of [9] that $f(d) = \Omega\left(C_{(V, \pi, f)}(d)\right) = \Omega\left(\mathbf{node}_T(d)\right)$. This establishes that, for each $d \in V$, we can determine the value of $f(d)$ from the labeled tree T, which is what we needed to prove.

3.2 (λ, k)-Simplification of a Component Tree

Component trees are sensitive to noise and other inaccuracies in the volume that can alter the structure of the tree. For example, many small component tree nodes may result from noise, and other nodes may represent structures that are of no interest (e.g., vitrified ice used in cryo-EM). In [8], a methodology called (λ, k)-simplification is presented that can be used to eliminate many irrelevant or erroneous parts of the tree while preserving the essential structural information it contains. An additional benefit of this simplification is that it reduces the size of the tree, so that the simplified tree can be more efficiently manipulated and analyzed than the original tree and is also easier to explore interactively. In order to describe the steps of (λ, k)-simplification in more detail, we introduce some additional notation.

Let T be a rooted tree. We use **Nodes**(T) to denote the nodes of T and **Children**$_T(C)$ to denote the set of all the children of C in T. A node C of T is said to be *critical* if $|\text{\bf Children}_T(C)| \neq 1$; the set of all critical nodes of T is denoted by **Crit**(T). We write T^{crit} to denote the rooted tree such that (i) **Nodes**$(T^{\text{crit}}) = \text{\bf Crit}(T) \cup \{\text{\bf root}(T)\}$, and (ii) for all C and D in **Nodes**(T^{crit}), C is an ancestor of D in T^{crit} if and only if C is an ancestor of D in T.

Given a component tree T_0, (λ, k)-simplification of T_0 can be understood as a three-step process that produces a simplified tree T_3:

- Step 1: Prune T_0 by removing nodes of size $\leq k$, giving a tree T_1.
- Step 2: Prune T_1 by removing branches of length $\leq \lambda$, giving a tree T_2.
- Step 3: Eliminate internal edges of length $\leq \lambda$ from T_2^{crit}, giving the output tree T_3.

To illustrate the effect of (λ, k)-simplification, we use a component tree of a density map of a wild-type procapsid of the bacteriophage $\phi 6$ [17]. A central slice of the density map and its component tree are shown in Figs. 3(a) and (b), respectively.

Step 1 - Removing Small Components: Step 1 of (λ, k)-simplification prunes the tree by removing nodes of small cardinality: All nodes that contain fewer than $k+1$ spels are removed. If k is suitably chosen, many nodes that result from noise in the digital picture will be removed by this step. Fig. 3(c) shows the tree produced by applying simplification step 1 with $k = 800$ to the component tree in Fig. 3(b).

Step 2 - Pruning Away Short Branches: Step 2 of (λ, k)-simplification prunes the tree by removing short branches. This pruning is applied to the result of step 1. Roughly speaking, the effect of this step is to remove all those leaves of the tree for which the difference between the label of the leaf and the label of its closest critical proper ancestor is less than or equal to λ. Fig. 3(d) shows the tree produced by applying simplification step 2 with $\lambda = 2$ to the tree in Fig. 3(c).

Step 3 - Elimination of Non-critical Nodes and Short Internal Edges: The result of step 2 is used as the input tree for step 3, which is the last step of (λ, k)-simplification. Roughly speaking, the effects of step 3 are to remove all non-critical nodes (with the exception of the root, which is not removed even if it is non-critical) and to also remove those critical nodes for which the difference between the label of that critical node and the label of its closest critical proper ancestor is $\leq \lambda$. The nodes that remain are the nodes of the final simplified tree; a node C of the final tree is an ancestor in that tree of a node D just if C was an ancestor of D in the original unsimplified tree.

We refer the reader to Sections 2.4 and 2.5 of [8] for a precise specification of steps 2 and 3, and a suggested way to implement those steps. Just as there are no formal criteria

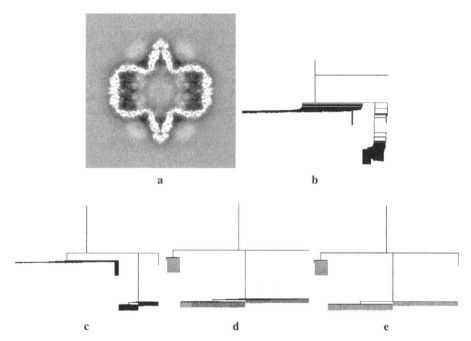

Fig. 3. Effect of (λ, k)-Simplification on a Component Tree of a Wild-Type $\phi6$ PC. (a) A central slice of a density map of a wild-type procapsid of the bacteriophage $\phi6$ [17]. (b) A component tree of the density map. Trees (c), (d), and (e) are respectively the results of applying step 1, steps 1 and 2, and all three steps of (λ, k)-simplification to the tree in (b). In this example we used $\lambda = 2$ and $k = 800$.

for selecting the parameters "level" and "width" (equivalent to brightness and contrast) when displaying a digital biomedical image on a computer screen (they are interactively adjusted until the display provides the user with the sought-after information), there are no formal criteria for selecting the parameters "λ" and "k"; they are determined by trying various combinations of values until the resulting tree structure reveals the sought-after biological information.

Fig. 3(e) shows the final output of $(2, 800)$-simplification when the original unsimplified tree is the tree in Fig. 3(b).

4   Using Component Trees to Distinguish Different Versions of a Bacteriophage Procapsid

In this section we present our initial findings from interactive exploration of component trees of procapsids of four versions of bacteriophage procapsid $\phi6$ using software that allows users to select tree nodes and display visualizations of the corresponding components [18]. Our findings suggest that suitably simplified component trees can be used to distinguish the four versions of the bacteriophage. This work also suggests that component trees may well be useful for identification of substructures in a macromolecular complex.

Fig. 4. Approximate Positions of the Proteins in a Wild-Type $\phi 6$ PC. The pink, blue, red, orange, and green arrows respectively indicate the approximate positions of P1a, P1b, P2, P4, and P7 proteins in a central slice of a wild-type $\phi 6$ procapsid.

Bacteriophage $\phi 6$ is a virus containing a genome of three segments of double-stranded RNA. The RNA packaging, replication, and transcription mechanisms of $\phi 6$ are very similar to those of reoviruses that contain species infectious to many animals, making the species an excellent model system for these important pathogens [11]. The initial step in $\phi 6$ replication is the assembly of a closed and unexpanded procapsid (PC) that is responsible for viral RNA packaging, transcription, and genome replication.

The PC has dodecahedral morphology with deeply recessed vertices; its diameter is approximately 860 Å. For wild-type (i.e., containing all its native protein elements) specimens of $\phi 6$ the PC comprises four proteins: P1, P2, P4, and P7. P1 is the structural protein responsible for the overall structure of the PC, which consists of 120 copies of this protein in two different conformations: P1a and P1b. The first, P1a, creates pentamers that are centered on the 12 faces of the PC, which therefore contains a total of 60 P1a proteins. The second, P1b, creates trimers at the vertices of the PC's dodecahedral skeleton; as there are 20 vertices, these trimers also contain a total of 60 proteins.

In wild-type specimens of $\phi 6$, each P2 protein is bound to the inner surface of the procapsid at a site close to a 3-fold symmetry axis of the dodecahedral structure [16], while each P4 is assembled on the procapsid's outer surface and overlies one of the pentamer faces. The last protein, P7, is the least characterized protein in the PC and its precise location is still a source of debate [11,17]. Cryo-EM studies of the related cystovirus indicate that each P7 is assembled into the PC at a site close to a 3-fold axis and the P1 shell [11]. Fig. 4 shows the approximate positions of the proteins in a PC: The pink, blue, red, orange, and green arrows show the approximate positions of P1a, P1b, P2, P4, and P7 proteins, respectively.

To illustrate how a visualization tool based on component trees can be used to explore biological structures, we now describe an investigation of the differences between component trees of density maps of four different recombinant procapsids of $\phi 6$—the wild-type version, and three versions obtained by genetic deletion in which one or both of the proteins P2 and P7 is missing. The four density maps used here originate from the work of Nemecek *et al.* [17] and are publicly available in the EMDataBank [5]. These density maps were obtained by single particle reconstruction from electron micrographs [6].

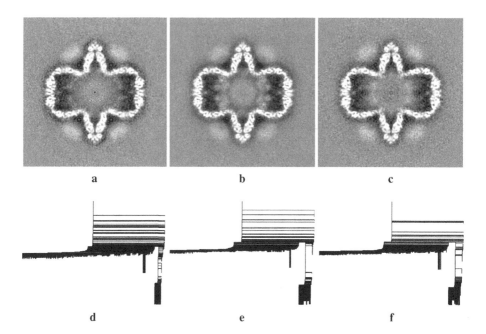

Fig. 5. Central Slices and Component Trees of Non-Wild-Type $\phi6$ PCs. Central slices of (a) PC14 (PC lacking P2 and P7); (b) PC124 (PC lacking P7); (c) PC147 (PC lacking P2). Component trees of (d) PC14; (e) PC124; (f) PC147.

One of the four density maps is of a PC of a wild-type $\phi6$, which consists of the four proteins P1, P2, P4, and P7; it will accordingly be referred to as PC1247. Another of the density maps is of a PC of a genetically modified $\phi6$ that was assembled from genes deleted for the sequence needed to synthesize protein P2. This PC therefore contains just the three proteins P1, P4, and P7, and will be referred to as P147. A third density map is of a $\phi6$ PC that contains P1, P2, and P4 but is missing the protein P7; this PC will be referred to as PC124. The last density map is of a $\phi6$ PC that will be referred to as PC14, because it contains P1 and P4 but contains neither of the proteins P7 and P2.

Each density map is a digital picture (V, π, f) in which V is a $400 \times 400 \times 400$ array of voxels. The graylevels in the original density maps (from EMDataBank) were floating point values, but we simplified the density maps by quantizing the graylevels to a set of just 256 equally spaced values represented by the integers $0, \ldots, 255$, where 0 corresponds to the minimum and 255 the maximum graylevel in the original density map. Central slices and component trees of the simplified density maps are shown in Figs. 3 and 5. For the component tree of the PC14 density map (see Fig. 5(d)), we found that (λ, k)-simplification with parameters $\lambda = 2$ and $k = 800$ yielded an appropriately simplified component tree. This simplified tree is shown in Fig. 6(a). The node indicated by the black arrow is the parent of 12 leaves that correspond to 12 P1a pentamers, and is the grandparent of 60 leaves that correspond to 60 P1b proteins which form the 20 P1b trimers. Copies of P4, the other protein in PC14, are represented by the 12 leaves in the orange oval on the left side of the component tree. Fig. 6(b) depicts surface

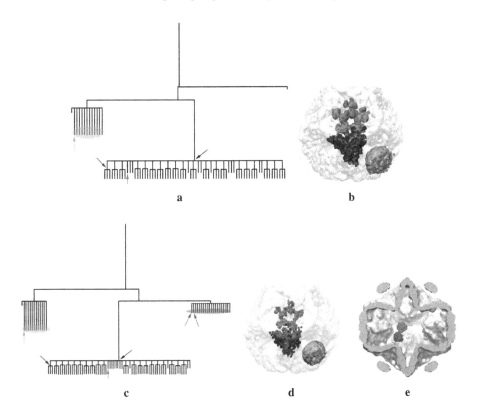

Fig. 6. PC14 and PC124 Procapsids and Their (2,800)-Simplified Component Trees. (a) The (2,800)-simplification of the PC14 component tree in Fig. 5(d); nodes associated with a P1a pentamer, a P1b trimer, and a P4 protein are indicated by the pink, blue, and orange arrows, respectively. (b) Surface renderings of the P1a pentamer (pink), P1b trimer (blue), and P4 protein (orange) represented by the nodes indicated by the colored arrows in (a). (c) The (2,800)-simplification of the PC124 component tree in Fig. 5(e). (d) Surface renderings of the nodes indicated by the pink, blue, and orange arrows in (c), which represent a P1a pentamer (pink), P1b trimer (blue), and P4 protein (orange). (e) Surface renderings of the P2 proteins represented by the two nodes indicated by the red arrows in (c).

renderings of the components associated with the nodes indicated by the colored arrows in (a), which represent a P1a pentamer (pink), P1b trimer (blue), and P4 protein (orange).

Thus Fig. 6 identifies the parts of the simplified component tree that represent the three substructures (i.e., the P1a pentamers, P1b trimers, and P4 proteins) which make up the PC14 procapsid.

The second component tree we investigated was the component tree of the PC124 density map. Fig. 6(c) shows its (2,800)-simplification. Note that, with the exception of the nodes in the red region, the component tree in Fig. 6(c) has the same structure as the (2,800)-simplified PC14 component tree in Fig. 6(a). (Again, the node indicated by the black arrow is the parent of 12 leaves that correspond to 12 P1a pentamers and 20 nodes (each with 3 leaf-children) that correspond to 20 P1b trimers, and is a "cousin" of

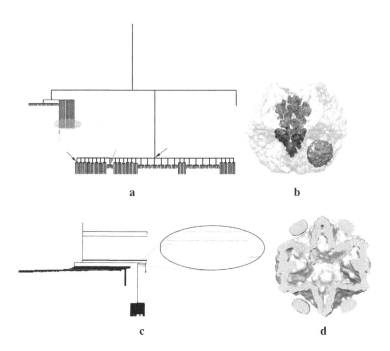

Fig. 7. Identifying Proteins in PC147 Component Trees. (a) The $(2,800)$-simplification and (c) the $(2,50)$-simplification of the PC147 component tree in Fig. 5(f). (b) Surface renderings of components in the $(2,800)$-simplified tree in panel (a) that represent a P1a pentamer (pink), a P1b trimer (blue), and a P4 protein (orange). (d) Surface renderings of the three components indicated by the green arrows in the $(2,50)$-simplified tree in panel (c), which represent P7 proteins.

12 leaves that represent 12 P4 proteins.) Copies of P2, the additional protein in PC124, are represented by the 20 leaves in the red oval on the right side of the component tree. Fig. 6(d) shows surface renderings of the P1a pentamers, P1b trimers, and P4 proteins associated with the nodes in Fig. 6(c) that are indicated by the pink, blue, and orange arrows respectively. The P2 proteins are on the inside of the PC shell; surface renderings of the nodes associated with two of the P2 proteins (indicated by the red arrows in Fig. 6(c)) are shown in Fig. 6(e).

In contrast to the simplified PC14 and PC124 component trees in Figs. 6(a) and 6(c), the $(2,800)$-simplified PC147 component tree, shown in Fig. 7(a), has a leaf for each P1a protein rather than a leaf for each P1a pentamer: The node indicated by the black arrow in Fig. 7(a) is still the parent of 12 nodes associated with P1a pentamers and 20 nodes associated with P1b trimers, but each of the nodes associated with the P1a pentamers now has five children—one for each P1a protein in the pentamer. The presence of a leaf for each P1a protein can be explained by a property of PC147 that is discussed in [17]: the very close proximity of each P1a to a P7. The densities from these two proteins are likely to be combined, resulting in components that are large enough to survive step 1 of $(2,800)$-simplification.

The proximity of each P1a to a P7 in PC147 also explains the absence of nodes that represent just the P7 proteins in Fig. 7(a). But when we used a smaller parameter value

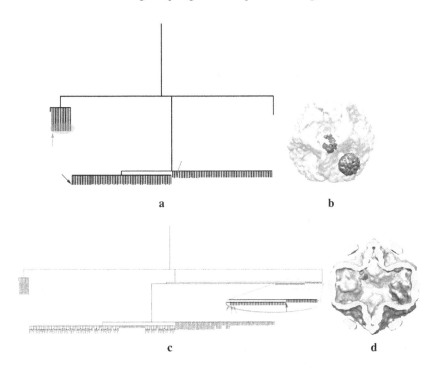

Fig. 8. Identifying Proteins in PC1247 Component Trees. (a) The $(2,800)$-simplification of the component tree presented in Fig. 8(e). (b) Surface renderings of the components associated with a P1a protein (pink), a P1b protein (blue), and a P4 protein (orange). (c) The $(2,300)$-simplification of the component tree presented in Fig. 8(e). (d) Surface renderings of the components indicated by the green (P7) and red (P2) arrows in (c).

$k = 50$ (instead of $k = 800$) in step 1 of the simplification process we found leaves that represent P7 proteins in the simplified component tree: In Fig. 7(c), which shows the $(2,50)$-simplified PC147 tree, these leaves can be seen in the enlarged part of the tree.

Surface renderings of the nodes of the $(2,800)$-simplified PC147 tree that are indicated by the pink (P1a pentamer), blue (P1b trimer), and orange (P4 protein) arrows in Fig. 7(a) are shown in Fig. 7(b). Surface renderings of the nodes of the $(2,50)$-simplified PC147 tree that are indicated by the green arrows in Fig. 7(c) are shown in Fig. 7(d); as mentioned above, these components correspond to P7 proteins.

The last density map we investigated was that of the wild-type procapsid PC1247. Fig. 8(a) depicts the $(2,800)$-simplification of the PC1247 component tree in Fig. 3(b). There is a node that is the parent of 60 leaves that correspond to 60 P1a proteins and the grandparent of 60 leaves that correspond to 60 P1b proteins; and there are 12 leaves, associated with P4 proteins, which are in the orange oval on the left side of the tree.

The $(2,800)$-simplified PC1247 component tree in Fig. 8(a) does not show nodes associated with the P7 and P2 proteins. However, the parameter choices $\lambda = 2$ and $k = 300$ yield a (λ,k)-simplified component tree in which nodes corresponding to the P2 and the P7 proteins are easily identified. The $(2,300)$-simplified component tree is shown in Fig. 8(c); the enlarged part of the tree shows the nodes associated with the P2 proteins (red shading) and the P7 proteins (green shading).

Surface renderings of the components associated with the nodes indicated by the pink (P1a pentamer), blue (P1b trimer), and orange (P4 protein) arrows in Fig. 8(a) are shown in Fig. 8(b). Surface renderings of the components associated with the nodes indicated by the red (P2 protein) and green (P7 protein) arrows in Fig. 8(c) are shown in Fig. 8(d).

5   Conclusion

We have discussed how component trees can be used to represent structural information in three-dimensional images (density maps) of biological macromolecular assemblies. Using a visualization tool based on component trees, we are able to interactively explore the structure of such biological specimens.

For four different versions of a bacteriophage procapsid, we have seen how the procapsid's protein substructures correspond to particular nodes of simplified component trees. We have also seen how component trees can provide an indication that certain substructures occur in very close proximity to each other.

Our work suggests that component trees are useful in investigating biological structures associated with microbes—in particular, virus particles. The authors are hopeful that future studies based on other density maps will provide further confirmation of this.

Acknowledgment. The work presented here is currently supported by the National Science Foundation (award number DMS-1114901) and the National Institute of General Medical Science (award number SC1 GM092781-01).

References

1. Caldairou, B., Naegel, B., Passat, N.: Segmentation of Complex Images Based on Component-Trees: Methodological Tools. In: Wilkinson, M.H.F., Roerdink, J.B.T.M. (eds.) ISMM 2009. LNCS, vol. 5720, pp. 171–180. Springer, Heidelberg (2009)
2. Carr, H., Snoeyink, J., Axen, U.: Computing Contour Trees in All Dimensions. Comput. Geom. 24, 75–94 (2003)
3. Dokládal, P., Bloch, I., Couprie, M., Ruijters, D., Urtasun, R., Garnero, L.: Topologically Controlled Segmentation of 3D Magnetic Resonance Images of the Head by Using Morphological Operators. Pattern Recogn. 36, 2463–2478 (2003)
4. Edelsbrunner, H., Harer, J.: Computational Topology: An Introduction. Am. Math. Soc., Providence (2010)
5. EMDataBank, http://emdatabank.org
6. Frank, J.: Three-Dimensional Electron Microscopy of Macromolecular Assemblies: Visualization of Biological Molecules in Their Native State. Oxford University Press, New York (2006)
7. Herman, G.T.: Geometry of Digital Spaces. Birkhäuser, Boston (1998)
8. Herman, G.T., Kong, T.Y., Oliveira, L.M.: Provably Robust Simplification of Component trees of Multidimensional Images. In: Brimkov, V.E., Barneva, R.P. (eds.) Digital Geometry Algorithms: Theoretical Foundations and Applications to Computational Imaging, pp. 27–69. Springer, Dordrecht (2012)

9. Herman, G.T., Kong, T.Y., Oliveira, L.M.: Tree Representation of Digital Picture Embeddings. J. Vis. Commun. Image. Represent. 23, 883–891 (2012)
10. Jones, R.: Connected Filtering and Segmentation Using Component Trees. Comput. Vis. Image Und. 75, 215–228 (1999)
11. Katz, G., Wei, H., Alimova, A., Katz, A., Morgan, D.G., Gottlieb, P.: Protein P7 of the Cystovirus $\phi 6$ is Located at the Three-Fold Axis of the Unexpanded Procapsid. PLoS ONE 7, e47489 (2012), doi:10.1371/journal.pone
12. Lawson, C.L., et al.: EMDataBank.org: Unified Data Resource for CryoEM. Nucl. Acids Res. 39, D456–D464 (2011)
13. Naegel, B., Passat, N., Boch, N., Kocher, M.: Segmentation Using Vector-Attribute Filters: Methodology and Application to Dermatological Imaging. In: Banon, G.J.F., Barrera, J., Braga-Neto, U. de, M., Hirata, N.S.T. (eds.) ISMM 2007: 8th International Symposium on Mathematical Morphology, vol. 1, pp. 239–250. INPE, São José dos Campos (2007)
14. Naegel, B., Wendling, L.: Combining Shape Descriptors and Component-Tree for Recognition of Ancient Graphical Drop Caps. In: Ranchordas, A., Araújo, H. (eds.) VISAPP 2009: 4th International Conference on Computer Vision Theory and Applications, vol. 2, pp. 297–302. INSTICC Press, Setúbal (2009)
15. Najman, L., Couprie, M.: Building the Component Tree in Quasi-Linear Time. IEEE Trans. Image Process. 15, 3531–3539 (2006)
16. Nemecek, D., Heymann, J.B., Qiao, J., Mindich, L., Steven, A.C.: Cryo-Electron Tomography of Bacteriophage $\phi 6$ Procapsids Shows Random Occupancy of the Binding Sites for RNA Polymerase and Packaging NTPase. J. Struct. Biol. 171, 389–396 (2010)
17. Nemecek, D., Qiao, J., Mindich, L., Steven, A.C., Heymann, J.B.: Packaging Accessory Protein P7 and Polymerase P2 Have Mutually Occluding Binding Sites inside the Bacteriophage $\phi 6$ Procapsid. J. Virol. 86, 11616–11624 (2012)
18. Oliveira, L.M., Kong, T.Y., Herman, G.T.: Using Component Trees to Explore Biological Structures. In: Herman, G.T., Frank, J. (eds.) Computational Methods for Three-Dimensional Microscopy Reconstruction, pp. 221–256. Birkhäuser, Boston (2014)
19. Ouzounis, G.K., Wilkinson, M.H.F.: Mask-Based Second-Generation Connectivity and Attribute Filters. IEEE Trans. Pattern Anal. Mach. Intell. 29, 990–1004 (2007)
20. Passat, N., Naegel, B., Rousseau, F., Koob, M., Dietemann, J.L.: Interactive Segmentation Based on Component-Trees. Pattern Recogn. 44, 2539–2554 (2011)
21. Salembier, P., Oliveras, A., Garrido, L.: Antiextensive Connected Operators for Image and Sequence Processing. IEEE Trans. Image Process. 7, 555–570 (1998)
22. Sarioz, D., Kong, T.Y., Herman, G.T.: History Trees as Descriptors of Macromolecular Structures. In: Bebis, G., et al. (eds.) ISVC 2006. LNCS, vol. 4291, pp. 263–272. Springer, Heidelberg (2006)
23. Tarjan, R.E.: Efficiency of a Good but Not Linear Set Union Algorithm. J. ACM 22, 215–225 (1975)
24. Wilkinson, M.H.F., Westenberg, M.A.: Shape Preserving Filament Enhancement Filtering. In: Niessen, W., Viergever, M. (eds.) MICCAI 2001. LNCS, vol. 2208, pp. 770–777. Springer, Heidelberg (2001)

A Combinatorial Technique for Construction of Triangular Covers of Digital Objects

Barnali Das[1], Mousumi Dutt[1], Arindam Biswas[1],
Partha Bhowmick[2,*], and Bhargab B. Bhattacharya[3]

[1] Department of Information Technology,
Bengal Engineering and Science University, Shibpur, Howrah, India
{bdbarnalidas,duttmousumi,barindam}@gmail.com
[2] Department of Computer Science and Engineering,
Indian Institute of Technology, Kharagpur, India
bhowmick@gmail.com
[3] Advanced Computing and Microelectronics Unit,
Indian Statistical Institute, Kolkata, India
bhargab@isical.ac.in

Abstract. The construction of a minimum-area geometric cover of a digital object is important in many fields of image analysis and computer vision. We propose here the first algorithm for constructing a minimum-area polygonal cover of a 2D digital object as perceived on a uniform triangular grid. The polygonal cover is triangular in the sense that its boundary consists of a sequence of edges on the underlying grid. The proposed algorithm is based on certain combinatorial properties of a digital object on a grid, and it computes the tightest cover in time linear in perimeter of the object. We present experimental results to demonstrate the efficacy, robustness, and versatility of the algorithm, and they indicate that the runtime varies inversely with the grid size.

Keywords: Triangular grid, Triangular polygon, Triangular cover, Polygonal cover, Shape analysis.

1 Introduction

Optimal polygonal covers of digital objects find diverse applications in many fields of image analysis and computer vision. Such covers are often obtained with a certain approximation parameter or with respect to an underlying grid that determines their precision. The grid is usually defined or characterized by the type of its constituent cells, which may be axis-parallel squares [3], equilateral triangles [8], or isosceles right-angled triangles [24]. Whether polygonal covers can efficiently be generated in low-order polynomial time for different grid patterns is a challenging problem, and this paper is focused on the specific problem when the cells comprising the underlying grid are equilateral triangles.

---

* Corresponding author.

R.P. Barneva, V.E. Brimkov, and J. Šlapal (Eds.): IWCIA 2014, LNCS 8466, pp. 76–90, 2014.
© Springer International Publishing Switzerland 2014

Fig. 1. A digital object (left) and its outer triangular covers (shown in red) for $g = 12$ (middle) and $g = 14$ (right)

A *triangular grid*, also called an *isometric grid* [10], is a grid formed by tiling the plane regularly with equilateral triangles. Given a triangular grid, the *triangular cover* of a digital object placed on the grid corresponds to a collection of equilateral triangles. Such a cover captures the structural and topological information of the concerned object, and finds useful applications in many fields, such as digital image processing and image analysis [19], geometric modeling [16], clustering and pattern analysis [3,16], ecological simulation [2] etc.

Triangular grids and hexagonal grids are duals of each other [19]. There exists a multitude of work in digital geometry in the framework of square grids, as well as triangular and hexagonal grids. Based on algorithmic comparison, it has been shown in [26] that there is a strong degree of similarity between square grids and hexagonal grids. This is also evidenced from the literature today, as we see several interesting work on characterization and construction of geometric primitives on triangular and hexagonal grids. Some of these related to digital lines, digital circles, and polygons, may be found in [9,12,18,22,23].

We propose here the first algorithm for construction of triangular covers of a digital object for different grid sizes, which is based on certain combinatorial rules. See Fig. 1 for triangular covers produced by our algorithm on a typical digital object for two different grid sizes. Although there exist similar work in the literature, no definite algorithm is available till date for the aforesaid problem. For example, the work in [24] is on tiling an image by isosceles right-angled triangles of the same size. Some investigation was made on analyzing the cellular topology with triangular grids, e.g., [21], where coordinate triplets are used to address the triangle pixels of both orientations, the edges between them, and the points at the triangle corners. The main advantage of a topological description is that it provides more information compared to the usual description, storing not only the pixels of the two-dimensional image but also the lower dimensional segments, the edges, and the points separating the pixels. To represent the points of a triangular grid and their neighborhood relations, coordinate triplets can be used, as shown in [19].

There also exist algorithms to compute the shortest distance between two given points on the triangular grid [17], as well as on the hexagonal grid [15], based on the distance function and the neighboring relations. In [20], it has been shown how distance computation can be done first with neighborhood sequence in 3D digital space and then by injection of the triangular grid to the cubic grid. For computing isoperimetrically optimal polygons in the triangular grid with Jordan-type neighborhood on the boundary, a method has been proposed in [22].

Some results on tiling in triangular grid have also been reported in recent time. There is no polynomial-time algorithm to solve the finite tiling problem. So, an attempt has been made in [25] to solve the problem using group theory. Such group-theoretic approaches for tiling a given finite region using polyominoes from triangular, square, or hexagonal lattices can be traced back to 1990's [7]. In [11], regular production systems are used as a tool for analyzing triangle tilings. In [1], a solution has been provided for the problem of N-tiling of a triangle ABC by a triangle T (tile) such that ABC can be represented as the union of N triangles congruent to T, overlapping only at their boundaries. In [14], a technique has been proposed for regular and irregular tiling of convex polygons with congruent triangles. Another technique proposed in [5] shows how to compute the locations of the vertices of a polygon, so that it can be perfectly tiled by lattice triangles. Tiling has also been implemented with other techniques such as using trichromatic colored-edged triangles, Golden Triangles, and Penrose Rhombs [4,6].

For construction of optimum-area outer and inner covers of a digital object against a given isothetic or a square grid, an algorithm is proposed in [3]. The algorithm proposed here for the triangular grid is essentially different from the one in [3] mainly for three reasons. First, the combinatorial cases on the triangular grid are different from those on the square grid. Second, the grid lines are non-rectilinear in a triangular grid, thus posing a higher level of difficulty compared to the square grid. Third, all the computations in the square grid can be restricted to integers only by setting the grid size to an integer value, whereas, the computations are in real domain in the case of triangular grid. Note that triangular mesh has also wide applications in the area of surface representation and construction. However, efficient computation of the tightest cover of a digital object on a triangular grid remains an open problem.

The paper is organized as follows. Definitions related to the triangular cover are stated in Sec. 2. Combinatorial rules that lead to the covering algorithm, with necessary explanations and runtime complexity, are discussed in Sec. 3. Experimental results on some datasets are presented in Sec. 4. Concluding notes and future scope of the work are given in Sec. 5.

2   Definitions

A *(digital) object* is a finite subset of $\mathbb{Z}^2$ consisting of one or more $k(= 4 \ or \ 8)$-*connected components* [13]. In our work, we consider the object as a single 8-connected component. A *triangular grid* (henceforth simply referred as *grid*) $\mathbb{T} := (\mathbb{L}_0, \mathbb{L}_{60}, \mathbb{L}_{120})$ consists of three sets of parallel *grid lines*, which are inclined at 0^0, 60^0, and 120^0 w.r.t. x-axis. The grid lines in $\mathbb{L}_0, \mathbb{L}_{60}, \mathbb{L}_{120}$ correspond

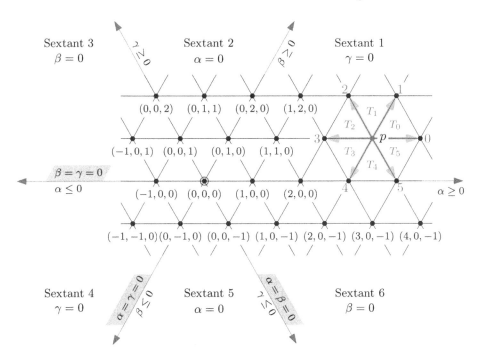

Fig. 2. Portion of a triangular grid, the UGTs $\{T_0, T_1, \ldots, T_5\}$ incident at a grid point p, and the direction codes $\{0, 1, \ldots, 5\}$ of neighboring grid points of p

to three distinct coordinates, namely α, β, γ. Three grid lines, one each from $\mathbb{L}_0$, $\mathbb{L}_{60}$, and $\mathbb{L}_{120}$, intersect at a (real) *grid point*. The distance between two consecutive *grid points* along a grid line is termed as *grid size*, g. A line segment of length g connecting two consecutive grid points on a grid line is called *grid edge*. The smallest-area triangle formed by three grid edges, one each from $\mathbb{L}_0$, $\mathbb{L}_{60}$, and $\mathbb{L}_{120}$, is called *unit grid triangle* (UGT). A portion of the triangular grid is shown in Fig. 2. It has six distinct regions called *sextants*, each of which is well-defined by two rays starting from $(0,0,0)$. For example, Sextant 1 is defined by the region lying between $\{\beta = \gamma = 0, \alpha \geq 0\}$ and $\{\alpha = \gamma = 0, \beta \geq 0\}$. One of α, β, γ is always 0 in a sextant. For example, $\gamma = 0$ in Sextant 1 and Sextant 4.

For a given grid point, p, there are six neighboring UGTs, given by $\{T_i : i = 0, 1, \ldots, 5\}$. The three coordinates of p are given by the corresponding moves along a/the shortest path from $(0,0,0)$ to p, measured in grid unit. For example, $(1,2,0)$ means a unit move along 0^0 followed by two unit moves along 60^0, starting from $(0,0,0)$. The grid point p can have six neighbor grid points, whose direction codes are given by $\{d' : i = 0, 1, \ldots, 5\}$.

A (finite) polygon P imposed on the grid $\mathbb{T}$ is termed as a *triangular polygon* if its sides are collinear with lines in $\mathbb{L}_0$, $\mathbb{L}_{60}$, and $\mathbb{L}_{120}$. It consists of a set of UGTs, and is represented by the (ordered) sequence of its vertices, which are grid points. It can also be represented by a sequence of directions and a start vertex. Its *interior* is defined as the set of points with integer coordinates lying

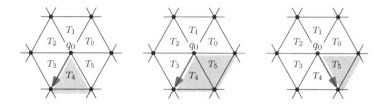

Fig. 3. Determining the start vertex, q_0, from the top-left object pixel, p_0, which would lie either in T_4 (left or middle configuration) or in T_5 (middle or right). Object-occupied UGTs are shown in gray, and start directions in red.

inside it. An *outer triangular polygon* is the minimum-area triangular polygon that tightly covers the digital object A in $\mathbb{T}$. Each UGT having at least one of its sides on the boundary of the outer polygon has object occupancy. An *outer triangular hole polygon* is the maximum-area triangular polygon that inscribes a hole or a concavity of A; that is, A lies outside the outer triangular hole polygon. The *outer triangular cover* (OTC), $\overline{P}$, is the set of outer triangular polygons and outer triangular hole polygons, such that the region given by the union of the outer triangular polygons minus the union of the interiors of the outer triangular hole polygons, contains a UGT if and only if it has object occupancy.

3   Obtaining Outer Triangular Cover

While obtaining the triangular cover $\overline{P}$ of A, our algorithm traverses the boundary of $\overline{P}$ with the invariant that A always lies left w.r.t. the direction of traversal. The top-left pixel on the boundary of the object, p_0, is given as input, from which the start vertex q_0 of $\overline{P}$ is computed, as shown in Fig. 3. If A contains any hole, then the top-left pixel of the hole boundary is also given as input. However, if the object contains a large concave region with 'narrow mouth', then the hole polygon inscribing that concave region is detected and traversed later from a UGT containing the border of the narrow mouth.

3.1   Classification of Vertices

The *object occupancy vector* corresponding to $\langle T_i : i = 0, 1, \ldots, 5\rangle$ incident at a grid point q is given by $A_q = \langle a_0 a_1 \cdots a_5\rangle$, where, for $i = 0, 1, \ldots, 5$, we have $a_i = 1$ if T_i incident at q has object occupancy, and 0 otherwise. The object occupancy of each T_i is determined by checking its interior integer points lying along its borders. The algorithm then uses the vector A_q to determine whether q lies on $\overline{P}$ and to compute the edges of $\overline{P}$ incident at q. For example, $A_q = 000011$ implies T_4 and T_5 are object-occupied. So, the direction of the incoming edge at q becomes 3 and that of the outgoing edge becomes 5, since the object lies left during traversal of $\overline{P}$.

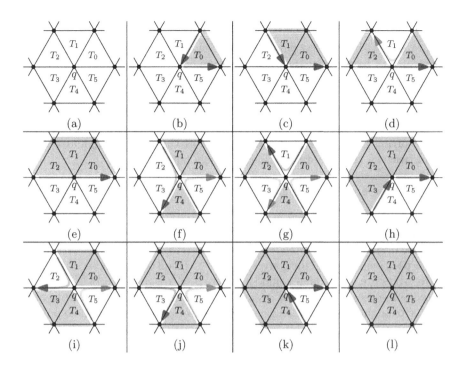

Fig. 4. Classification of a grid point q for (a) $k = 0$, (b) $k = 1$, (c, d) $k = 2$, (e-g) $k = 3$, (h-j) $k = 4$, (k) $k = 5$, (l) $k = 6$. A red line indicates the traversal direction of the outer triangular polygon, and a blue line indicates that of an outer triangular hole polygon. Detailed explanations are given in Sec. 3.1.

A grid point q lies on $\overline{P}$ if A_q contains both 0's and 1's. If $A_q = 0^6$, then q lies outside $\overline{P}$ (Fig. 4a); and $A_q = 1^6$ implies q lies strictly inside $\overline{P}$ (Fig. 4l). Apart from these two cases, there arise $2^6 - 2 = 62$ combinatorial cases of A_q. These 62 cases can be simplified to following 5 cases only, with their corresponding sub-cases (Fig. 4b-k), depending on the number of 1's in A_q, which is denoted by $k(= 1, 2, \ldots, 5)$.

• $k = 1$: There are 6 sub-cases depending on the position of 1-bit in A_q. For each sub-case, the angle at q is $\pi/3$ (Fig. 4b). If $a_i = 1$, then the *direction pair* representing the incoming and the outgoing edges at q is $((i + 4) \bmod 6, i)$.

• $k = 2$: Let $a_i = a_j = 1$. If $j = (i + 1) \bmod 6$, then there are 6 sub-cases, each with angle at q as $2\pi/3$ and direction pair $((i + 5) \bmod 6, i)$, as shown in Fig. 4c. Otherwise, there are 9 sub-cases, each with q occurring twice in $\overline{P}$—once (as a vertex or a non-vertex edge point) on the outer triangular polygon, and once on a triangular hole polygon (Fig. 4d).

• $k = 3$: Following are three possibilities.

- The positions of three 1-bits in A_q are i, $(i+1) \bmod 6$, and $(i+2) \bmod 6$. There are 6 sub-cases in which q is not a vertex but a grid point on an edge of $\overline{P}$, and with direction pair (i, i); see Fig. 4e.
- Two 1-bits have positions i and $(i+1) \bmod 6$, and the third 1-bit in any position j other than $(i+2) \bmod 6$. There are 12 sub-cases, each with q occurring twice in $\overline{P}$—once as a vertex of the outer triangular polygon, and once as a vertex of a triangular hole polygon (Fig. 4f).
- All three 1-bits in A_q are non-consecutive. There are 2 sub-cases, each with q occurring thrice in $\overline{P}$—once as a vertex of the outer triangular polygon, and twice as vertices of two triangular hole polygons (Fig. 4g).

- $k = 4$: There are three possibilities as follows.

  - The positions of four 1-bits in A_q are $(i+j) \bmod 6$, for $j = 0, 1, 2, 3$ (Fig. 4h). There are 6 sub-cases, each with angle at q as $4\pi/3$ and direction pair $((i+7) \bmod 6, i)$.
  - The positions of two 1-bits are i and $(i+1) \bmod 6$, and those of the other two 1-bits are j and $(j+1) \bmod 6$, where $j = (i+3) \bmod 6$ (Fig. 4i). There are 3 sub-cases, each with q occurring twice in $\overline{P}$—once as a vertex of the outer triangular polygon, and once as a vertex of a triangular hole polygon.
  - The positions of three 1-bits are i, $(i+1) \bmod 6$, and $(i+2) \bmod 6$, and that of the fourth 1-bit is $(i+4) \bmod 6$ (Fig. 4j). There are 6 sub-cases, each with q occurring twice in $\overline{P}$, as in the previous case.

- $k = 5$: There are 6 sub-cases depending on the position of 0-bit in A_q (Fig. 4k). For each sub-case, the angle at q is $5\pi/3$. If $a_i = 0$, then the direction pair representing the incoming and the outgoing edges at q is $((i+3) \bmod 6, (i+1) \bmod 6)$.

3.2 Determining the Next Direction

The outgoing direction d' at a grid point q lying on $\overline{P}$ is computed from the incoming direction d and $A_q := \langle a_0 a_1 \cdots a_5 \rangle$. It is based on the following observation: The incoming direction d implies $a_j = 1$ and $a_{(j+1) \bmod 6} = 0$, where $j = (d+2) \bmod 6$. Hence, j is incremented (in modulo 6 arithmetic) until the next 1-bit in A_q is encountered, say at j', so that $a_{j'} = 1$. The outgoing direction is then given by $d' = j'$. For example, in Fig. 4f, $A_q = 110010$ and $d = 5$ gives $j = (5+2) \bmod 6 = 1$, implying $a_1 = 1$ and $a_2 = 0$; since $a_3 = 0$ and $a_4 = 1$, we get $d' = 4$ for the outer triangular polygon. While traversing the hole polygon in Fig. 4f, $d = 2$ gives $j = (2+2) \bmod 6 = 4$, implying $a_4 = 1$ and $a_5 = 0$; since $a_0 = 1$, we get $d' = 0$.

3.3 Algorithm to Construct $\overline{P}$

Algorithm 1 (CONSTRUCT-OTC) shows the important steps to construct the outer triangular polygon of a digital object (single connected component without

---

Algorithm 1. CONSTRUCT-OTC

    Input   : $A, \mathbb{T}, p_0$
    Output: L
1 $q \leftarrow q_0 \leftarrow$ FIND-START(p_0)
2 $L \leftarrow \{q_0\}$
3 $d \leftarrow$ FIND-START-DIR(q_0)
4 **do**
5     $d' \leftarrow$ FIND-DIR(d, q)
6     $q \leftarrow$ NEXT-VERT(d', q)
7     $d \leftarrow d'$
8     $L \leftarrow L \cup \{q\}$
9 **while** $q \neq q_0$
10 **return** L

---

holes). It takes the object, A, the triangular grid, $\mathbb{T}$, and the top-left pixel of $p_0 \in A$, as input, and generates the sequence of vertices of $\overline{P}$ as output. The same algorithm is also used to construct the other polygons, e.g., outer triangular hole polygons and also other outer triangular polygons in case of a digital object with multiple components. For each component or each hole, the top-left pixel is supplied as an input.

In Step 1 of the algorithm, the top-left grid point (start vertex) of $\overline{P}$, namely q_0, is determined from p_0 using the procedure FIND-START. Its related details have already been explained in Sec. 3. The list L stores the sequence of vertices of $\overline{P}$. It is initialized with $\{q_0\}$ in Step 2. In Step 3, the start direction is determined based on the UGT containing p_0, as explained earlier and illustrated in Fig. 3. In the **do-while** loop (Steps 4–9), the outgoing direction d' and the next vertex are determined, until the next vertex coincides with the start vertex.

The outgoing direction d' at q is determined by the procedure FIND-DIR in Step 5. Its related operations are explained in Sec. 3.2. From d', the next vertex is easily computed by NEXT-VERT in Step 6, using $\mathbb{T}$ and the coordinates of q. The ordered list of vertices, L, is returned finally in Step 10.

3.4   Correctness and Time Complexity

From the combinatorial cases explained in Sec. 3.1 and from the strategy of getting outgoing directions explained in Sec. 3.2, we can make the following observation: For every grid edge e lying on $\overline{P}$, one UGT incident on e has object-occupancy and the other has not. Further, among these two UGTs, the interior of only the former lies inside $\overline{P}$. This implies that none of the UGTs whose edges lie on $\overline{P}$ and interiors lie inside $\overline{P}$, can be removed from $\overline{P}$, which proves the minimum-area criterion of $\overline{P}$ produced by Algorithm 1.

To prove that the start vertex q_0 is finally reached when the algorithm terminates, observe that the algorithm maintains the invariant that the object A always lies left during traversal. Since the incoming direction is left out while commencing the start direction from q_0, the outgoing direction from the last grid point on $\overline{P}$ finally leads to q_0, hence ensuring its successful termination.

For time complexity analysis, let n be the number of pixels comprising the contour (border) of a connected component comprising A. A UGT may contain as few as one contour pixel, but for every seven UGTs through which the contour passes in succession, there would be at least $O(g)$ contour pixels, as g is the grid size of $\mathbb{T}$. Hence, there are at most $O(n/g)$ UGTs containing the object contour, or, the number of grid points on $\overline{P}$ is at most $O(n/g)$.

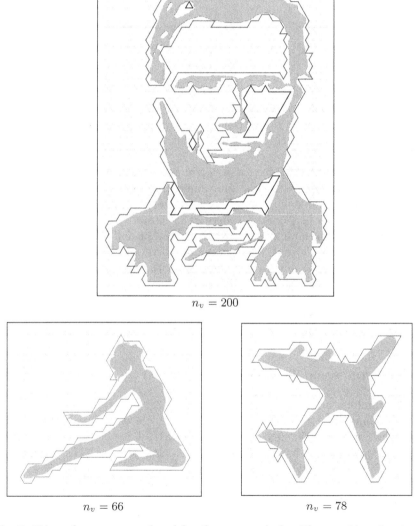

Fig. 5. Triangular covers produced by the proposed algorithm on `Lincoln`, `dancer`, and `aircraft` images, with $g = 20$

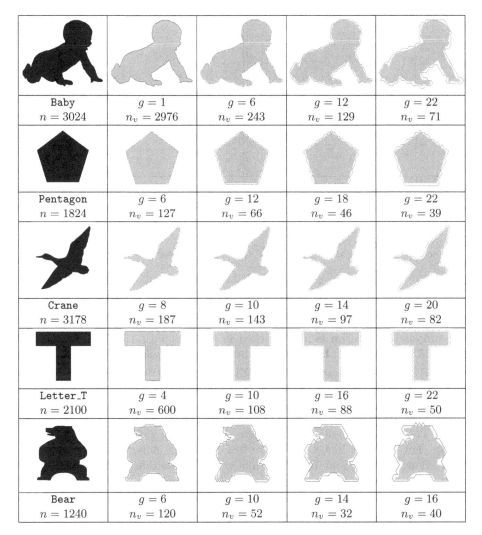

Baby $n = 3024$	$g = 1$ $n_v = 2976$	$g = 6$ $n_v = 243$	$g = 12$ $n_v = 129$	$g = 22$ $n_v = 71$
Pentagon $n = 1824$	$g = 6$ $n_v = 127$	$g = 12$ $n_v = 66$	$g = 18$ $n_v = 46$	$g = 22$ $n_v = 39$
Crane $n = 3178$	$g = 8$ $n_v = 187$	$g = 10$ $n_v = 143$	$g = 14$ $n_v = 97$	$g = 20$ $n_v = 82$
Letter_T $n = 2100$	$g = 4$ $n_v = 600$	$g = 10$ $n_v = 108$	$g = 16$ $n_v = 88$	$g = 22$ $n_v = 50$
Bear $n = 1240$	$g = 6$ $n_v = 120$	$g = 10$ $n_v = 52$	$g = 14$ $n_v = 32$	$g = 16$ $n_v = 40$

Fig. 6. Experimental results for different grid sizes. (n = number of pixels on the object contour; n_v = number of vertices of $\overline{P}$.)

To find the start vertex q_0 and the corresponding start direction, $O(1)$ time is required. So, Steps 1–3 of Algorithm 1 takes $O(1)$ time. Each UGT is checked in $O(g)$ time, as explained in Sec. 3.1. As each grid point has six neighboring UGTs, the time complexity to compute the object occupancy vector at each grid point q lying on $\overline{P}$ is $6 \cdot O(g) = O(g)$. Thus, for q, outgoing direction and next grid point finding are performed in $O(g)$ time. Hence, the *worst-case* time complexity of Algorithm 1 is $O(n/g) \cdot O(g) = O(n)$.

For the *best-case* analysis, observe that there can be $O(g^2)$ contour pixels in a UGT, which implies that there can be $O(n/g^2)$ UGTs containing the object contour, or, the number of grid points on $\overline{P}$ is $O(n/g^2)$. As each UGT can

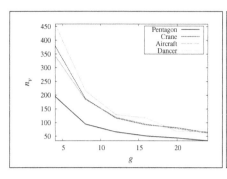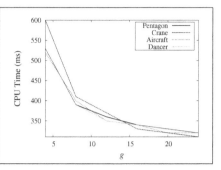

Fig. 7. Plots on (left) number of vertices and (right) CPU time versus g for `aircraft`, `dancer`, `Crane`, and `pentagon` images

be checked for object occupancy in $O(g)$ time, we get the time complexity as $O(n/g^2) \cdot O(g) = O(n/g)$.

4   Experimental Results

The proposed algorithm is implemented in C in Ubuntu 12.04, 32-bit, kernel version 3.5.0-43-generic, the processor being Intel core i5-3570, 3.40GHz FSB. The algorithm has been used to construct triangular covers of digital objects in various binary images of logo datasets, real-world objects, and geometric shapes, for different grid sizes.

The outer triangular covers (OTCs) of three different objects are shown in Fig. 5. The outer triangular polygons are shown in red and outer hole polygons in blue. For `Lincoln` image, there are five outer hole polygons along with the outer triangular polygon. It may be noticed that there are a few small holes in the top-left region of the digital object of `Lincoln` image; however, only one outer hole polygon in that region is reported by our algorithm, as the other holes are not large enough to accommodate any polygon for $g = 20$.

Figure 6 shows the outer triangular covers of five objects from our datasets. The second and the fourth ones are geometric shapes, the third one is a logo image, and the first and the fifth are real-world objects. In each row, the input object, the number of pixels constituting its border (n), the OTCs for some grid sizes, and the number of vertices (n_v) of the OTCs are shown. The results indicate that n_v decreases with the increase of g. As g decreases, smaller irregularities on the object border do not leave much impression on $\overline{P}$, but the gross shape of the object gets captured through fewer vertices. The cover $\overline{P}$ captures the structural property of the object for smaller values of g, and it loses information with increasing values of g. For higher values of g, output complexity is low, resulting in less storage, but topological information is not always retained. For example, some of the outer triangular hole polygons may disappear for higher values of grid size. Such a collection of $\overline{P}$ for changing grid size might be useful for multi-scale shape analysis.

Object		Isothetic Cover		Triangular Cover

Object	Area	Perimeter	Outer Cover	n_v	Area	Perimeter
Letter_T	117966	2100	Triangular	88	136014	2320
			Isothetic	8	140800	2080
Kangaroo	92292	4926	Triangular	128	113955	3392
			Isothetic	114	122112	3392
Crane	101362	2864	Triangular	92	118057	2512
			Isothetic	110	123904	2848

Fig. 8. Comparison of the triangular covers produced by the proposed algorithm with the isothetic covers produced by the algorithm in [3], all for $g = 16$. (n_v=number of vertices.)

Table 1 presents a consolidated information of the OTCs for several objects at different grid sizes. In all cases, the number of vertices decreases with g, which is also evident from the plots showing n_v versus g in Fig. 7. The CPU times plotted against different grid sizes for four images are shown in this figure. As evident from these plots, the CPU time taken to compute the OTCs falls sharply with the increasing value of g, which indicates that the runtime of the algorithm on a real-world image follows the best-case runtime complexity of $O(n/g)$, as explained in Sec. 3.4.

Table 1. Summary of experimental results

Object A	|A|	n	g	n_{60}	n_{120}	n_{240}	n_{300}	n_v	n_T	$\|\overline{P}\|$	$p_{\overline{P}}$	T
Baby	132629	3024	4	45	159	149	47	400	19760	136901	2744	600
			8	26	81	69	29	205	5093	141141	2760	410
			12	2	68	52	7	129	2350	146531	2568	360
			16	4	46	30	9	89	1348	149427	2592	340
			20	3	35	21	7	66	887	153633	2540	320
			24	2	31	19	5	57	619	154388	2472	310
Pentagon	116711	1636	4	19	81	77	18	195	17185	119061	1516	530
			8	10	40	36	9	95	4401	121964	1528	390
			12	0	36	30	0	66	2004	124957	1464	360
			16	2	26	22	1	51	1152	127701	1504	340
			20	3	22	16	3	44	745	129038	1540	330
			24	2	18	12	2	34	537	133936	1560	320
Aircraft	50607	2724	4	44	131	123	45	343	7835	54282	2460	520
			8	23	75	57	29	184	2086	57809	2496	400
			12	5	61	43	11	120	998	62229	2352	350
			16	5	46	38	6	95	593	65735	2320	340
			20	8	34	28	8	78	396	68589	2360	320
			24	4	30	24	4	62	286	71333	2208	320
Crane	101362	2864	4	47	147	135	50	379	15199	105302	2652	600
			8	21	77	65	24	187	3951	109493	2648	410
			12	5	57	49	6	117	1828	113983	2520	370
			16	2	49	35	6	92	1065	118057	2512	330
			20	4	40	34	4	82	704	121936	2560	320
			24	3	32	26	3	64	506	126204	2544	310
Letter_T	117966	2100	4	168	135	129	168	600	17480	121105	2800	570
			8	20	68	64	19	171	4502	124763	2288	410
			12	45	47	43	44	179	2057	128262	2676	360
			16	10	37	31	10	88	1227	136014	2320	330
			20	0	32	24	1	57	805	139430	2180	320
			24	1	26	20	1	48	551	137428	2184	310
Dancer	46280	2997	4	75	161	141	82	459	7295	50541	2772	530
			8	28	84	74	30	216	1967	54511	2608	390
			12	11	56	52	10	129	951	59298	2472	350
			16	16	46	38	17	117	565	62631	2592	340
			20	4	33	27	4	68	387	67030	2280	330
			24	8	29	23	8	68	274	68340	2352	320
Bear	36450	1238	4	27	67	53	31	178	5511	38181	1208	130
			8	7	39	25	11	82	1443	39990	1144	90
			12	4	24	12	7	47	690	43024	1116	90
			16	7	15	13	5	40	395	43786	1136	80
			20	1	20	10	3	34	273	47285	1100	70
			24	3	18	8	5	34	189	47139	1176	70

$|A|$ = number of object pixels; n = number of pixels on the boundary of A; g = grid size; n_θ = number of vertices with internal angle θ (= 60, 120, 240, 300); n_v = total number of vertices of $\overline{P}$; n_T = number of UGTs in outer cover; $|\overline{P}|$ = number of pixels in $\overline{P}$; T = CPU time in milliseconds; $p_{\overline{P}}$ = perimeter of $\overline{P}$.

To estimate the effectiveness of triangular covers in capturing the object shape, a comparison of the proposed algorithm has been done with the algorithm for finding isothetic covers proposed in [3]. Figure 8 shows this comparison for a geometric shape (Letter_T), a logo image (Kangaroo), and a real-world object (Crane), all for $g = 16$.

The object in Letter_T is perfectly aligned with the horizontal and the vertical lines of the isothetic grid. Hence, its isothetic cover has fewer vertices compared to its triangular cover; however, in area measure, the triangular cover is better compared to the isothetic cover. For the image Kangaroo, the number of vertices of the triangular cover is more than that of the isothetic cover; but the area of the triangular cover is comparatively less, indicating that it is tighter than the isothetic cover, and hence more compact. The boundary of the object in Crane has some portions aligned roughly along the grid lines of the triangular grid, which results in fewer vertices of the triangular cover compared to the isothetic cover; and here also, the area measure of the triangular cover is less than that of the isothetic cover. Thus, the triangular covers are particularly effective, both in terms of vertex count and area measure, when the object boundary has some alignment with the underlying grid. An isothetic cover, in general, may have a smaller set of vertices, and so its computational time may be less. On the other hand, a triangular cover usually requires less area to encompass the object; and hence, it offers more compactness, although it can have more vertices and may require more computational time.

5   Conclusion

We have proposed an algorithm for constructing a minimum-area cover of a digital object on a triangular grid. It can be used to determine all outer triangular polygons including the outer triangular hole polygons in a digital object with multiple components and holes. The minimum-area cover provides a compact geometric information of the object, which is useful in shape characterization along with those obtained from orthogonal or isothetic covers. Further, its runtime can be controlled by tuning the grid size, which offers a mechanism of trading off speed with granularity of the object boundary. Experimental results demonstrate that a triangular cover can serve as an approximate estimate of shape complexity and of similar other signatures used for characterizing a digital object.

Acknowledgments. The second and the third authors gratefully acknowledge the respective supports provided by INSPIRE Program under Department of Science and Technology, Government of India, and Major Project Scheme under University Grant Commission, Government of India.

References

1. Beeson, M.: Triangle tiling I: The tile is similar to ABC or has a right angle. arXiv preprint arXiv:1206.2231 (2012)
2. Birch, C.P.D., Oom, S.P., Beecham, J.A.: Rectangular and hexagonal grids used for observation, experiment and simulation in ecology. Ecological Modelling 206(3), 347–359 (2007)
3. Biswas, A., Bhowmick, P., Bhattacharya, B.B.: Construction of isothetic covers of a digital object: A combinatorial approach. Journal of Visual Communication and Image Representation 21(4), 295–310 (2010)

4. Bodini, O., Rémila, E.: Tilings with trichromatic colored-edges triangles. Theoretical Computer Science 319(1), 59–70 (2004)
5. Butler, S., Chung, F., Graham, R., Laczkovich, M.: Tiling polygons with lattice triangles. Discrete & Computational Geometry 44(4), 896–903 (2010)
6. Clason, R.G.: Tiling with golden triangles and the penrose rhombs using logo. Journal of Computers in Mathematics and Science Teaching 9(2), 41–53 (1989)
7. Conway, J.H., Lagarias, J.C.: Tiling with polyominoes and combinatorial group theory. Journal of Combinatorial Theory, Series A 53(2), 183–208 (1990)
8. Daniel, H., Tom, K., Elmar, L.: Exploring simple triangular and hexagonal grid polygons online. arXiv preprint arXiv:1012.5253 (2010)
9. Freeman, H.: Algorithm for generating a digital straight line on a triangular grid. IEEE Transactions on Computers 100(2), 150–152 (1979)
10. Gardner, M.: Knotted Doughnuts and Other Mathematical Entertainments. Freeman and Company, New York (1986)
11. Goodman-Strauss, C.: Regular production systems and triangle tilings. Theoretical Computer Science 410(16), 1534–1549 (2009)
12. Innchyn, H.: Geometric transformations on the hexagonal grid. IEEE Transactions on Image Processing 4(9), 1213–1222 (1995)
13. Klette, R., Rosenfeld, A.: Digital Geometry: Geometric Methods for Picture Analysis. Morgan Kaufmann, San Francisco (2004)
14. Laczkovich, M.: Tilings of convex polygons with congruent triangles. Discrete & Computational Geometry 48(2), 330–372 (2012)
15. Luczak, E., Rosenfeld, A.: Distance on a hexagonal grid. IEEE Transactions on Computers 25(5), 532–533 (1976)
16. Nagy, B.: Neighbourhood sequences in different grids. Ph.D. thesis, University of Debrecen (2003)
17. Nagy, B.: Shortest paths in triangular grids with neighbourhood sequences. Journal of Computing and Information Technology 11(2), 111–122 (2003)
18. Nagy, B.: Characterization of digital circles in triangular grid. Pattern Recognition Letters 25(11), 1231–1242 (2004)
19. Nagy, B.: Generalised triangular grids in digital geometry. Acta Mathematica Academiae Paedagogicae Nyíregyháziensis 20(1), 63–78 (2004)
20. Nagy, B.: Distances with neighbourhood sequences in cubic and triangular grids. Pattern Recognition Letters 28(1), 99–109 (2007)
21. Nagy, B.: Cellular topology on the triangular grid. In: Barneva, R.P., Brimkov, V.E., Aggarwal, J.K. (eds.) IWCIA 2012. LNCS, vol. 7655, pp. 143–153. Springer, Heidelberg (2012)
22. Nagy, B., Barczi, K.: Isoperimetrically optimal polygons in the triangular grid with Jordan-type neighbourhood on the boundary. International Journal of Computer Mathematics 90(8), 1–24 (2012)
23. Shimizu, K.: Algorithm for generating a digital circle on a triangular grid. Computer Graphics and Image Processing 15(4), 401–402 (1981)
24. Subramanian, K.G., Wiederhold, P.: Generative models for pictures tiled by triangles. Science and Technology 15(3), 246–265 (2012)
25. Sury, B.: Group theory and tiling problems. Symmetry: A Multi-Disciplinary Perspective 16(16), 97–117 (2011)
26. Wüthrich, C.A., Stucki, P.: An algorithmic comparison between square-and hexagonal-based grids. CVGIP: Graphical Models and Image Processing 53(4), 324–339 (1991)

Equivalent 2D Sequential and Parallel Thinning Algorithms

Kálmán Palágyi

Department of Image Processing and Computer Graphics,
University of Szeged, Hungary
palagyi@inf.u-szeged.hu

Abstract. Thinning is a frequently applied skeletonization technique: border points that satisfy certain topological and geometric constraints are deleted in iteration steps. Sequential thinning algorithms may alter just one point at a time, while parallel algorithms can delete a set of border points simultaneously. Two thinning algorithms are said to be equivalent if they can produce the same result for each input binary picture. This work shows that the existing 2D fully parallel thinning algorithm proposed by Manzanera et al. is equivalent to a topology-preserving sequential thinning algorithm with the same deletion rule.

Keywords: Discrete Geometry, Digital Topology, Thinning, Equivalent Thinning Algorithms.

1 Introduction

Thinning is an iterative layer-by-layer erosion until only some skeleton-like shape features are left [6,12]. Thinning algorithms use *reduction operator*s that transform binary pictures (i.e., images containing only black and white points) only by changing some black points to white ones, which is referred to as deletion.

Parallel thinning algorithms are comprised of *reduction*s that can delete a set of border points simultaneously [2,7,13], while sequential thinning algorithms traverse the boundary of objects and may remove just one point at a time [7,13]. In the parallel case the initial set of black points is considered when the deletion rule is evaluated. On the contrary, the set of black points is dynamically altered during a sequential reduction.

Thinning algorithms generally classify the set of black points into two (disjoint) subsets: the deletion rule of an algorithm is evaluated for the elements of its set of *interesting* points, and black points in its *constraint set* are not taken into consideration. Constraint sets comprise the set of interior points (i.e., black points that are not border points) [6] and they may contain some types of border points in subiteration-based (or directional) parallel algorithms [2] or border points that are not in the active subfield in the case of subfield-based parallel algorithms [2]. In addition, endpoints (i.e., some border points that provide important geometrical information relative to the shape of the objects [2])

R.P. Barneva, V.E. Brimkov, and J. Šlapal (Eds.): IWCIA 2014, LNCS 8466, pp. 91–100, 2014.
© Springer International Publishing Switzerland 2014

or isthmuses (i.e., generalization of curve and surface interior points [1]) can also be accumulated in the constraint sets.

Two reductions (i.e., thinning phases) are said to be *equivalent* if they produce the same result for each input binary picture [9]. A deletion rule is called *equivalent* if it yields a pair of equivalent parallel and sequential reductions [9]. As far as we know, no one showed that there exists a pair of equivalent parallel and sequential thinning algorithms.

The sequential approach suffers from the drawback that different visiting order of interesting points may yield various results. Order-independent sequential reductions can produce the same result for any visiting order of the elements in the actual set of interesting points [3,10]. It is obvious that only an order-independent sequential reduction can be equivalent to a parallel one.

In [9] the author gave some sufficient conditions for equivalent deletion rules. This paper shows that the deletion rule of the known 2D fully parallel thinning algorithm proposed by Manzanera et al. [8] is equivalent. Hence an example of a pair of equivalent parallel an sequential thinning algorithms is presented. In addition the topological correctness of these algorithms is also proved.

The rest of this paper is organized as follows. Section 2 gives an outline from basic notions and results from digital topology, topology preservation, and equivalent reductions. Then in Section 3 we rephrase the known parallel thinning algorithm proposed by Manzanera et al. [8]. In Section 4 we show that the considered parallel algorithm is equivalent to a topology-preserving sequential thinning algorithm. Hence the topological correctness of an existing parallel thinning algorithm is also verified. Finally, we round off the paper with some concluding remarks.

2   Basic Notions and Results

We use the fundamental concepts of digital topology as reviewed by Kong and Rosenfeld [6].

Let p be a point in the 2-dimensional digital space $\mathbb{Z}^2$. Let us denote $N_m(p)$ the set of points that are m-*adjacent* $(m = 4, 8)$, see Fig. 1. Note that, throughout this paper, all figures are depicted on the square grid that is dual to $\mathbb{Z}^2$.

Fig. 1. The considered adjacency relations in $\mathbb{Z}^2$. The set $N_4(p)$ contains point p and the four points marked "◆". The set $N_8(p)$ contains $N_4(p)$ and the four points marked "◇".

The equivalence classes relative to the m-*connectivity* relation (i.e., the transitive closure of the reflexive and symmetric m-adjacency relations) are the m-*components* of a set of points $X \subseteq \mathbb{Z}^2$.

A $(8,4)$ *digital picture* $\mathcal{P}$ is a quadruple $(\mathbb{Z}^2, 8, 4, B)$. Each element of $\mathbb{Z}^2$ is said to be a *point* of $\mathcal{P}$. Each point in $B \subseteq \mathbb{Z}^2$ is called a *black point*. Each point in $\mathbb{Z}^2 \setminus B$ is said to be a *white point*. A *black component* is an 8–component of B, while a *white component* is a 4–component of $\mathbb{Z}^2 \setminus B$.

A black point is called a *border point* in a $(8,4)$ picture if it is 4–adjacent to at least one white point. A black point in a picture is said to be an *interior point* if it is not a border point.

A reduction (on a 2D picture) is *topology-preserving* if each black component (as a set of points) in the original picture contains exactly one black component of the produced picture, and each white component in the output picture contains exactly one white component of the input picture [6].

A black point is *simple* in a picture if and only if its deletion is a topology-preserving reduction [6]. We mention now the following characterization of simple points:

Theorem 1. [6] *Black point p is simple in a picture $(\mathbb{Z}^2, 8, 4, B)$ if and only if all of the following conditions hold:*

1. $N_8(p) \setminus \{p\}$ *contains exactly one black component.*
2. p *is a border point.*

Recall that a deletion rule is equivalent if it yields a pair of equivalent parallel and (order-independent) sequential reductions. The author gave the following sufficient conditions for equivalent deletion rules:

Theorem 2. [9] *Let R be a deletion rule. Let $(\mathbb{Z}^2, 8, 4, B)$ be an arbitrary picture, and let $q \in B$ be any point that is deleted from that picture by R. Deletion rule R is equivalent if the following two conditions hold for any $p \in B \setminus \{q\}$:*

1. *If p can be deleted from picture $(\mathbb{Z}^2, 8, 4, B)$ by R, then p can be deleted from picture $(\mathbb{Z}^2, 8, 4, B \setminus \{q\})$ by R.*
2. *If p cannot be deleted from picture $(\mathbb{Z}^2, 8, 4, B)$ by R, then p cannot be deleted from picture $(\mathbb{Z}^2, 8, 4, B \setminus \{q\})$ by R.*

Reductions associated with parallel thinning phases may delete a set of black points and not just a single simple point. Hence we need to consider what is meant by topology preservation when a number of black points are deleted simultaneously. Various authors proposed sufficient conditions for reductions to preserve topology [4,5,11]. The author established the following ones:

Theorem 3. [9] *A (parallel) reduction with deletion rule R is topology-preserving if the following conditions hold:*

1. R *is equivalent.*
2. R *deletes only simple points.*

3   An Existing Fully Parallel 2D Thinning Algorithm

In this section we recall the 2D parallel thinning algorithm proposed by Manzanera et al. [8]. That existing algorithm falls into the category of fully parallel thinning [2] since it uses the same reduction in each thinning phase (i.e., iteration step). The set of interesting points associated with the reduction of the algorithm contains the set of border points in the actual picture. Hence its constraint set comprises all interior points in the input picture of the actual iteration step.

Manzanera et al. [8] gave the deletion rule of their algorithm by three classes of patterns. The base patterns α_1, α_2, and β are depicted in Fig. 2. All their rotated versions are patterns as well, where the rotation angles are 90°, 180°, and 270°. All α_1- and α_2-type of elements are *removing patterns*, while β and its rotated versions are the *preserving patterns*. A black point is designated to be deleted if at least one removing pattern matches it, but it is not matched by any preserving pattern.

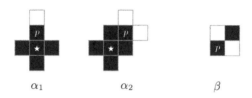

α_1 α_2 β

Fig. 2. The three base patterns associated with the 2D fully parallel algorithm proposed by Manzanera et al. Notations: each black position is a black point; each white element is a white point; black positions marked "p" are the central positions of the patterns. (Note that each position marked "★" is an interior point.)

In order to prove that the considered parallel algorithm is equivalent to a sequential thinning algorithm, we rephrase its deletion rule $\mathcal{MBPL}$ by eliminating the preserving patterns. The rephrased rule is given by the set of 32 removing patterns $\mathcal{P} = \{P_1, \ldots, P_{32}\}$, see Fig. 3.

It can be readily seen that the 16 patterns $\{P_1, \ldots, P_{16}\}$ are associated with the removing patterns of type α_1 with respect to the preserving patterns of type β. Similarly, the remaining 16 patterns $\{P_{17}, \ldots, P_{32}\}$ are assigned to the removing patterns of type α_2 with respect to the preserving patterns of type β.

One iteration step of the considered 2D fully parallel algorithm is sketched by Algorithm 1. The constraint set C contains all interior points of the actual binary image, hence deletion rule $\mathcal{MBPL}$ is evaluated for the set of border points $Y = X \setminus C$. A border point $p \in Y$ is deletable (i.e., $\mathcal{MBPL}\,(p, X) = \mathbf{true}$) if at least one pattern depicted in Fig. 3 matches it. Deletable points (i.e., elements of the set D) are removed simultaneously. The entire reduction (i.e., iteration step) is repeated until no points are deleted (i.e., $D = \emptyset$).

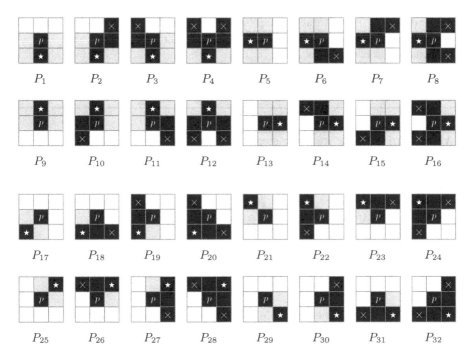

Fig. 3. The set of 32 patterns $\mathcal{P} = \{P_1, \ldots, P_{32}\}$ associated with the deletion rule $\mathcal{MBPL}$ (see Algorithm 1). Notations: each black position is a black point; each white element is a white point; each (don't care) position depicted in grey matches either a black or a white point; "p" indicates the central point of a pattern; each position marked "★" is an interior point; each position marked "×" is a border point (i.e., it is not an interior point).

4 A Sequential 2D Thinning Algorithm Being Equivalent to a Parallel One

We propose a sequential thinning algorithm that uses the deletion rule $\mathcal{MBPL}$ (see Algorithm 1). One iteration step of the derived algorithm is given by Algorithm 2. It (i.e., a phase of the thinning process) is repeated until stability is reached.

The derived sequential algorithm removes the actually visited border point $p \in Y$ right away if it is deletable in the actual image (i.e., $\mathcal{MBPL}(p, X) =$ **true**), therefore the set of black points X is dynamically altered within an iteration step.

We show that the new sequential algorithm is equivalent to the considered parallel thinning algorithm (i.e., Algorithms 1 and 2 are equivalent). In order to prove it, let us state some properties of the deletion rule $\mathcal{MBPL}$ (see Fig. 3). For the sake of brevity, a black point is said to be *deletable*, if it can be deleted by $\mathcal{MBPL}$ (i.e., at least one pattern in $\mathcal{P}$ matches it).

Algorithm 1.   one iteration step of the fully parallel thinning algorithm

---

Input: set of black points X and set of interior points $C \subset X$
Output: set of black points X
$Y = X \setminus C$
$D = \{ \, p \mid p \in Y$ and $\mathcal{MBPL}\,(p, X) = \textbf{true} \, \}$
$X = X \setminus D$

Algorithm 2.   one iteration step of the sequential thinning algorithm

---

Input: set of black points X and set of interior points $C \subset X$
Output: set of black points X
$Y = X \setminus C$
foreach $p \in Y$ **do**
   | **if** $\mathcal{MBPL}\,(p, X) = \textbf{true}$ **then**
   | | $X = X \setminus \{p\}$

end

Proposition 1. *Each deletable point is 8-adjacent to at least one interior point.*

Notice that each 3×3 pattern in $\mathcal{P}$ contains a position marked "★" (i.e., an element that is an interior point), see Fig. 3.

Proposition 2. *In each pattern in $\mathcal{P}$, the "opposite" positions of the "★" elements are white points.*

Figure 4 illustrates the "opposite" white points that are associated with an interior point (marked "★"). Since each deletable point is 8-adjacent to at least one interior point by Proposition 1, Proposition 2 holds.

Proposition 3. *All deletable points are simple points.*

It is really apparent that both conditions of Theorem 1 hold from a careful examination of the patterns in $\mathcal{P}$.

Proposition 4. *Deletable points are not interior points.*

Condition 2 of Theorem 1 is not satisfied for interior points. Hence they are not deletable points by Proposition 3.

Proposition 5. *All the 26 patterns in $\{P_3, P_5, \ldots, P_{16}, P_{19}, P_{21}, \ldots, P_{32}\}$ can be derived from the six base patterns in $\{P_1, P_2, P_4, P_{17}, P_{18}, P_{20}\}$.*

Pattern P_3 is a reflected version of P_2; P_{19} is a reflected version of P_{18}; P_5, P_9, and P_{13} are the rotated versions of P_1; P_6, P_{10}, and P_{14} are the rotated versions of P_2; P_7, P_{11}, and P_{15} are the rotated versions of P_3; P_8, P_{12}, and P_{16} are the rotated versions of P_4; P_{21}, P_{25}, and P_{29} are the rotated versions of P_{17}; P_{22}, P_{26}, and P_{30} are the rotated versions of P_{18}; P_{23}, P_{27}, and P_{31} are the rotated

Fig. 4. Interior points marked "★" that are 8-adjacent to the central point "p". The depicted white points are the "opposite" positions associated with the corresponding interior point.

versions of P_{19}; and P_{24}, P_{28}, and P_{32} are the rotated versions of P_{20}. Hence Proposition 5 holds.

Now we are ready to state the key theorem.

Theorem 4. *Deletion rule $\mathcal{MBPL}$ is equivalent.*

Proof. Let $(\mathbb{Z}^2, 8, 4, B)$ be an arbitrary picture. To prove this theorem we must show that both conditions of Theorem 2 are satisfied. Let $q \in B$ be an arbitrary deletable point. Then the following two points are to be proved for any point $p \in B \setminus \{q\}$:

1. If p is a deletable point in picture $(\mathbb{Z}^2, 8, 4, B)$, then p is a deletable point in picture $(\mathbb{Z}^2, 8, 4, B \setminus \{q\})$.
2. If p is not a deletable point in picture $(\mathbb{Z}^2, 8, 4, B)$, then p is not a deletable point in picture $(\mathbb{Z}^2, 8, 4, B \setminus \{q\})$.

Since deletion rule $\mathcal{MBPL}$ is given by 3×3 patterns, there is nothing to prove if $q \notin N_8(p)$. Hence it is sufficient to show that if we alter just one element in any pattern $P \in \mathcal{P}$, then

- we get another pattern $P' \in \mathcal{P}$ (written as $P \Rightarrow P'$),
- we do not get a pattern in $\mathcal{P}$ (i.e., $P \Rightarrow P'' \notin \mathcal{P}$), and the considered black position (or the altered white pattern element in question) is not a deletable point, or
- point p is in the constraint set (hence p is not a deletable point before/after the deletion of q).

Let us see the six base patterns of $\mathcal{P}$, see Fig. 5. It is easy to see that we do not need to take "don't care" positions (depicted in grey) into account. Note that pattern elements marked "★" (i.e., interior points) cannot be altered by Proposition 4.

Let us consider the remaining black and white elements of the six base patterns (see Fig. 5).

- Let white point a (in base patterns P_1, P_2, P_{17}, and P_{18}) be the deleted point q.
  - If the point marked "◇" (a "don't care" position in that patterns that is 8-adjacent to the point a in question) is white, then we do not get a pattern in $\mathcal{P}$, and $a = q$ is not a deletable point, since no deletable points are matched by a β-type preserving pattern (see Fig. 2).

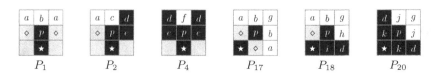

Fig. 5. The six base patterns in $\mathcal{P}$. Note that positions marked "★" are interior points, and black points marked "d" are border points (i.e., they are not interior points).

- If the point marked "◇" is black, then we get another pattern in $\mathcal{P}$: $P_1 \Rightarrow P_2$ or P_3; $P_2 \Rightarrow P_4$; $P_{17} \Rightarrow P_{18}$ or P_{19}; $P_{18} \Rightarrow P_{20}$.
- Let white point b (in base patterns P_1, P_{17}, and P_{18}) be the deleted point q. Then we do not get a pattern in $\mathcal{P}$, and $b = q$ is not a deletable point by Propositions 1 and 2.
- Let white point c (in base pattern P_2) be the deleted point q. Since $c = q$ is a deleted point, c is 8-adjacent to an interior point by Proposition 1. It can be readily seen that black point e may be the only interior point that is 8-adjacent to c.
  - If the point marked "◇" (a "don't care" position in P_2) is black, then p is an interior point before deletion of q. Hence p is in the constraint set, and it is also not a deletable point after the deletion of q.
  - If the point marked "◇" is white, then we get another pattern in $\mathcal{P}$: $P_2 \Rightarrow P_{13}$ or P_{15}.
- Let black point d (in base patterns P_2, P_4, P_{18}, and P_{20}) be the deletable point q. If $d = q$ is deleted, then we get another pattern in $\mathcal{P}$: $P_2 \Rightarrow P_1$; $P_4 \Rightarrow P_2$ or P_3; $P_{18} \Rightarrow P_{17}$; $P_{20} \Rightarrow P_{18}$ or P_{19}.
- Let black point e (in base patterns P_2 and P_4) be deletable point q. Then we do not get a pattern in $\mathcal{P}$, and $e = q$ is not a deletable point by Propositions 1 and 2.
- Let white point f (in base pattern P_4) be the deleted point q. Then we do not get a pattern in $\mathcal{P}$, and $f = q$ is not a deletable point by Propositions 1 and 2. (Elsewise, p is an interior point before deletion of q. Hence p is in the constraint set, and it is also not a deletable point after the deletion of q.)
- Let white point g (in base pattern P_{18}, P_{19}, and P_{20}) be the deleted point q. Then we do not get a pattern in $\mathcal{P}$, and $g = q$ is not a deletable point, since no deletable points are matched by a β-type preserving pattern (see Fig. 2).
- Let white point h (in base pattern P_{18}) be the deleted point q. Since $h = q$ is a deleted point, it is 8-adjacent to an interior point by Proposition 1. It can be readily seen that black point i may be the only interior point that is 8-adjacent to h. Then we get another pattern in $\mathcal{P}$: $P_{18} \Rightarrow P_1$.
- Let black point i (in base patterns P_{18}) be the deletable point q. Then we do not get a pattern in $\mathcal{P}$, and $i = q$ is not a deletable point by Propositions 1 and 2.
- Let white point j (in base pattern P_{20}) be the deleted point q. Since $j = q$ is a deleted point, it is 8-adjacent to an interior point by Proposition 1. It can be readily seen that a black point k may be the only interior point that is 8-adjacent to c. In this case we get another pattern in $\mathcal{P}$: $P_{20} \Rightarrow P_3$ or P_6.

– Let black point k (in base pattern P_{20}) be the deletable point q. Then we do not get a pattern in $\mathcal{P}$, and $k = q$ is not a deletable point by Propositions 1 and 2.

Since the remaining 26 patterns in P_{20} can be derived from the six base patterns by Proposition 5, the proof can be carried out for all elements of the set of patterns $\mathcal{P}$ (that is associated with the deletion rule $\mathcal{MBPL}$). $\square$

Theorem 4 means that the 2D fully parallel thinning algorithm proposed by Manzanera et al. [8] (see Algorithm 1) and the sequential algorithm with the same deletion rule $\mathcal{MBPL}$ (see Algorithm 2) are equivalent.

An easy consequence of Theorem 4 and Proposition 3 is that both algorithms (i.e., the original parallel and the derived sequential ones) are topology-preserving, hence we presented an alternative proof concerning the topological correctness of an existing parallel thinning algorithm.

5   Conclusions

In an earlier work the author laid a bridge between the parallel and the sequential reductions. Some sufficient conditions for equivalent parallel and order-independent sequential reductions with the same deletion rule were given.

This work shows that an existing 2D (fully parallel) thinning algorithm is equivalent to a topology-preserving sequential thinning algorithm. Hence an example is found that a useful parallel algorithm can be replaced by a sequential one.

Acknowledgements. This work was supported by the European Union and co-funded by the European Social Fund. Project title: "Telemedicine-focused research activities on the field of Mathematics, Informatics and Medical sciences." Project number: TÁMOP-4.2.2.A-11/1/KONV-2012-0073.

References

1. Bertrand, G., Couprie, M.: Transformations topologiques discrètes. In: Coeurjolly, D., Montanvert, A., Chassery, J. (eds.) Géométrie Discrète et Images Numériques, pp. 187–209. Hermès Science Publications (2007)
2. Hall, R.W.: Parallel connectivity-preserving thinning algorithms. In: Kong, T.Y., Rosenfeld, A. (eds.) Topological Algorithms for Digital Image Processing, pp. 145–179. Elsevier Science B.V. (1996)
3. Kardos, P., Palágyi, K.: Order-independent sequential thinning in arbitrary dimensions. In: Proc. Int. Conf. Signal and Image Processing and Applications, SIPA 2011, pp. 129–134 (2011)
4. Kardos, P., Palágyi, K.: On Topology Preservation in Triangular, Square, and Hexagonal Grids. In: Proc. 8th Int. Symposium on Image and Signal Processing and Analysis, ISPA 2013, pp. 782–787 (2013)

5. Kong, T.Y.: On topology preservation in 2–d and 3–d thinning. International Journal of Pattern Recognition and Artificial Intelligence 9, 813–844 (1995)
6. Kong, T.Y., Rosenfeld, A.: Digital topology: Introduction and survey. Computer Vision, Graphics, and Image Processing 48, 357–393 (1989)
7. Lam, L., Lee, S.-W., Suen, S.-W.: Thinning methodologies — A comprehensive survey. IEEE Trans. Pattern Analysis and Machine Intelligence 14, 869–885 (1992)
8. Manzanera, A., Bernard, T.M., Pretêux, F., Longuet, B.: n-dimensional skeletonization: a unified mathematical framework. Journal of Electronic Imaging 11, 25–37 (2002)
9. Palágyi, K.: Deletion Rules for Equivalent Sequential and Parallel Reductions. In: Ruiz-Shulcloper, J., Sanniti di Baja, G. (eds.) CIARP 2013, Part I. LNCS, vol. 8258, pp. 17–24. Springer, Heidelberg (2013)
10. Ranwez, V., Soille, P.: Order independent homotopic thinning for binary and grey tone anchored skeletons. Pattern Recognition Letters 23, 687–702 (2002)
11. Ronse, C.: Minimal test patterns for connectivity preservation in parallel thinning algorithms for binary digital images. Discrete Applied Mathematics 21, 67–79 (1988)
12. Siddiqi, K., Pizer, S. (eds.): Medial representations – Mathematics, algorithms and applications. Computational Imaging and Vision, vol. 37. Springer (2008)
13. Suen, C.Y., Wang, P.S.P. (eds.): Thinning methodologies for pattern recognition. Series in Machine Perception and Artificial Intelligence, vol. 8. World Scientific (1994)

Sufficient Conditions for General 2D Operators to Preserve Topology

Péter Kardos and Kálmán Palágyi

Department of Image Processing and Computer Graphics,
University of Szeged, Hungary
{pkardos,palagyi}@inf.u-szeged.hu

Abstract. An important requirement for various applications of binary image processing is to preserve topology. This issue has been earlier studied for two special types of image operators, namely, reductions and additions, and there have been some sufficient conditions proposed for them. In this paper, as an extension of those earlier results, we give novel sufficient criteria for general operators working on 2D pictures.

Keywords: Digital Topology, Binary Operators, Topology Preservation.

1 Introduction

The term *digital topology* was first used by Rosenfeld [15, 16] and its subject is concerned with certain topological properties of sets in digital pictures, like adjacency and connectedness [1–3, 6, 10]. Binary pictures contain black or white elements, which are called *points*. We can make a distinction among three types of binary operators: *reductions* (or reductive operators [2]) which never change (delete) white elements to black ones, *additions* (or augmentative operators [2]) which never turn black elements into white ones, and general (reductive-augmentative [2]) operators which may modify both black and white elements.

One of the frequently researched issues in digital topology is topology preservation. As two examples for topology preserving procedures, we can mention thinning, which is an iterative object reduction technique for producing skeleton-like features [7, 11, 13, 18], and the generation of skeleton by influence zones (SKIZ), which serves as the discrete analog for Voronoi diagrams [17]. A further possible situation for the need of topological correct operators occurs when contour smoothing is applied with the aim of noise reduction in combination with thinning algorithms [12].

Ronse gave some sufficient conditions for topology-preserving 2D reductions [14], then Ma established analogous conditions to preserve topology by 3D reductions [9], while Kardos and Palágyi simplified those criteria and also proposed their equivalents for additions [8]. Although, they also formulated a simple relationship for general operators, which is based on their decomposition into a topology preserving reduction and a topology preserving addition, there are not yet any sufficient conditions, similar to the Ronse's ones, that would directly examine the point configurations changed by general transformations.

R.P. Barneva, V.E. Brimkov, and J. Šlapal (Eds.): IWCIA 2014, LNCS 8466, pp. 101–112, 2014.
© Springer International Publishing Switzerland 2014

Furthermore, sometimes it is not straightforward to find such a proper decomposition of operators which is required by the condition given in [8]. For instance, let us consider the pattern in Fig. 1(a), and the result of the right shift operator for that picture (see Fig. 1(b)), which clearly preserves the topology. Now let us trivially decompose that operator into a reduction that modifies each black point with a white left neighbor and an addition that changes each white pixel having a black left neighbor. We can observe in Figs. 1(c-d) that none of them preserve the topology, as the reduction removes each object (i.e., isolated black points) of the picture, and the addition merges those objects.

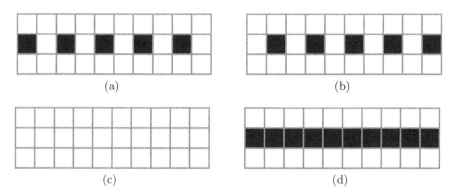

(a)

(b)

(c)

(d)

Fig. 1. Picture of a simple pattern (a), result of a right shift operator (b), and results for the reduction (c) and the addition (d) that we get if we trivially decompose the right shift operator

This motivated us for some further research in this field to deal with the sufficiency for topology preserving general operators. In this paper we introduce our results for 2D (8,4) pictures [6].

The rest of this paper is organized as follows. Section 2 gives a short review on the basic notions and results of digital topology. In Section 3 we introduce our sufficient conditions for topology preserving general operators, while in Section 4 we show an example for a 2D operator and, to illustrate the usefulness of our conditions, we prove its topological correctness by them. Finally, we round off the paper with some concluding remarks.

2   Basic Notions and Results

This section summarizes the fundamental concepts of digital topology as reviewed by Kong and Rosenfeld in [6].

The elements of the 2D digital space $\mathbb{Z}^2$ are called *points*. Note that, throughout this paper, all figures are depicted on the square grid that is dual to $\mathbb{Z}^2$. For a point $p \in \mathbb{Z}^2$, $N_j(p)$ refers to the set of points being *j-adjacent* to p, and let $N_j^*(p) = N_j(p) \setminus \{p\}$ $(j = 4, 8)$. A *unit square* contains four mutually 8-adjacent points, while a *small triangle* is formed by three mutually 8-adjacent points.

A small triangle T has exactly one corner point that is 4-adjacent to the remaining two points of that triangle.

We call a sequence of distinct points $\langle s_0, s_1, \ldots, s_m \rangle$ a j-*path* of length m from point s_0 to point s_m in a non-empty set of points S if each point of the sequence is in S and s_i is j-adjacent to s_{i-1} $(i = 1, 2, \ldots, m)$. Note that a single point is a j-path of length 0. Two points are said to be j-*connected* in the set S if there is a j-path in S between them A set of points S is j-*connected* in the set of points $Y \supseteq S$ if any two points in S are j-connected in Y.

A 2D $(8, 4)$ *binary digital picture* is a quadruple $(\mathbb{Z}^2, 8, 4, B)$ [6], where $\mathbb{Z}^2$ is the set of 2D discrete points. Each point in $B \in \mathbb{Z}^2$ is called a *black point* and has a value of 1 assigned to it. Each point in $\mathbb{Z}^2 \setminus B$ is called a *white point* and has a value of 0 assigned to it. An *object* is a maximal 8-connected set of black points, while a *white component* is a maximal 4-connected set of white points.

A *simple point* is a point whose color does not influence the topology of the picture [6]. We will make use of the following characterization of simple points:

Theorem 1. [1] *Point p is simple in picture $(\mathbb{Z}^2, 8, 4, B)$ if and only if all of the following conditions hold.*

1. $N_4^*(p) \setminus B \neq \emptyset$.
2. *The picture $(\mathbb{Z}^2, 8, 4, N_8^*(p) \cap B)$ contains exactly one object.*

Some examples of simple and non-simple points are shown in Fig. 2.

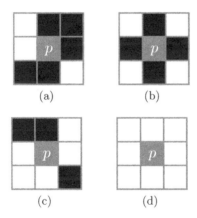

(a)   (b)

(c)   (d)

Fig. 2. Examples of a simple (black or white) point (a) and three non-simple points (b-d) in (8,4) pictures. Point p in (b) violates Condition 1 of Theorem 1, while in the remaining two cases, Condition 2 of Theorem 1 is not satisfied.

Here we extend the notions simple sets and minimal non-simple sets for of black points defined in [14] to general sets of points composed of black and white points. Let $\mathcal{P}$ be a picture, $\mathcal{O}$ be a general operator, and S be the set of all points in $\mathcal{P}$ that are modified by $\mathcal{O}$. We call S is as a *simple set*, if the elements of S can

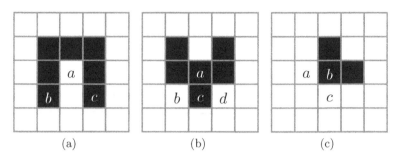

(a)   (b)   (c)

Fig. 3. Examples for a simple set (a) and two non-simple sets (b-c). Set $\{a, b, c\}$ in (a) is simple; set $\{a, b, c, d\}$ in (b) is non-simple but not minimally non-simple, as its proper subset $\{a, c\}$ is also non-simple; set $\{a, b, c\}$ is minimally non-simple, as all of its proper subsets (i.e., $\{a\}$, $\{b\}$, $\{c\}$, $\{a, b\}$, $\{a, c\}$, and $\{b, c\}$) are simple.

be arranged in a *simple sequence* $\langle s_1, \ldots, s_m \rangle$ such that s_1 is simple and each s_i is simple after the colors of points in the set $\{s_1, \ldots, s_{i-1}\}$ are modified in $\mathcal{P}$ ($i = 2, \ldots, m$). (By definition, let the empty set be simple.) In other words, there is an order of points in a simple set such that if we modify their color one by one, we get a sequence of topology preserving transformations. A set of points is *minimal non-simple* if it is not simple, but any of its proper subsets is simple.

For a better understanding on simple, non-simple, and minimal non-simple sets, Fig. 3 shows some examples.

Kong presented an important relationship for simple sets in the view of topology preserving reduction in [5], which result can be extended for general operators as stated in [8]:

Theorem 2. [8] *A general operator $\mathcal{O}$ is topology preserving if the set of points modified by $\mathcal{O}$ is simple for any picture.*

For (8,4) pictures, Ronse also proposed some sufficient conditions on topology preservation which examine point configurations:

Theorem 3. [9, 14] *A reduction $\mathcal{R}$ is topology preserving, if all of the following conditions hold for any picture $(\mathbb{Z}^2, 8, 4, B)$.*

1. *Only simple pixels are deleted by $\mathcal{R}$.*
2. *For any two 4-adjacent black pixels, $p, q \in B$ that are deleted by $\mathcal{R}$, p is simple in $(\mathbb{Z}^2, 8, 4, B \backslash \{q\})$, or q is simple in $(\mathbb{Z}^2, 8, 4, B \backslash \{p\})$.*
3. *$\mathcal{R}$ never deletes any object contained in a unit square.*

In [8] it was shown that Condition 3 of the previous theorem also examines some configurations that are excluded by Conditions 1 or 2, and that there is a simpler equivalent of Condition 2. Therefore the above result can be simplified to the following form:

Theorem 4. [8] *A reduction $\mathcal{R}$ is topology preserving, if all of the following conditions hold for any picture $(\mathbb{Z}^2, 8, 4, B)$.*

1. *Only simple points are deleted by $\mathcal{R}$.*
2. *For any two 4-adjacent black pixels, p, q that are deleted by $\mathcal{R}$, p is simple in $(\mathbb{Z}^2, 8, 4, B \backslash \{q\})$.*
3. *$\mathcal{R}$ never deletes any object contained in a unit square that has two 8-adjacent but not 4-adjacent black points.*

Kardos and Palágyi formulated a duality theorem between additions and reductions in [8], and by using that relationship, they established the following sufficient conditions for topology preserving additions:

Theorem 5. *[8] An addition $\mathcal{A}$ is topology preserving, if all of the following conditions hold in any picture $(\mathbb{Z}^2, 8, 4, B)$.*

1. *Only simple white points are modified by $\mathcal{A}$.*
2. *For any two 8-adjacent white points, p, q that are modified by $\mathcal{A}$, p is simple in picture $(\mathbb{Z}^2, 8, 4, B \cup \{q\})$.*

3 Sufficient Conditions for Topology Preserving General Operators

In this section we present some relationships on general 2D operators. As our main result, we introduce our sufficient conditions on topology preservation for the general case.

We note that the following two lemmas and the statement of Theorem 6 are extensions of the results in [4] and [14] on reductions.

As a preparation for the discussion of Lemma 1 we prove the following property.

Proposition 1. *If a non-simple set of points S is not minimal non-simple, then S has a minimal non-simple proper subset.*

If all the proper subsets of S would be simple, then by definition S would be minimal non-simple, hence S has at least one non-simple proper subset Q. If Q is not minimal non-simple, then again one of its proper subset is non-simple. This train of thoughts can be recursively continued at most until we get a proper non-simple subset containing only one point, which, by definition, is minimal non-simple.

According to the following lemma, in order to guarantee topology preservation it is sufficient to examine the point configurations that may consitute minimal non-simple sets.

Lemma 1. *A general operator is topology preserving, if it does not change any minimal non-simple set.*

Proof. We give an indirect proof. Let us assume that the condition of the lemma holds for $\mathcal{O}$ (i.e., $\mathcal{O}$ does not change any minimal non-simple set in any picture), yet $\mathcal{O}$ does not preserve the topology for a picture. This, by Theorem 2, is only possible if the points modified by $\mathcal{O}$ constitute a non-simple set, which we denote by S. S is not minimal non-simple by our assumptions. However, by Proposition 1, S has at least one minimal non-simple proper subset, which contradicts our initial assumption.   □

The next lemma determines the widest region of critical point configurations to examine for ensuring topology preservation. Thanks to this relationship, we can restrict our attention to a relatively small number of possible situations for our purpose.

Lemma 2. *Each minimal non-simple set of a binary picture* $\mathcal{P} = (\mathbb{Z}^2, 8, 4, B)$ *is contained in a unit square.*

Proof. Let X be a minimal non-simple set in $\mathcal{P}$. If X contains only one element, then the lemma trivially holds. Let us suppose that X contains at least two points. It is sufficient to prove that any two points $p, q \in X$ are 8-adjacent. As X is a minimal non-simple set, the sets $X \setminus \{p, q\}$ and $X \setminus \{q\}$ are simple. Let $X \setminus \{p, q\} = X_b \cup X_w$ where $X_b = (X \setminus \{p, q\}) \cap B$, while $X_w = (X \setminus \{p, q\}) \setminus B$. In picture $(\mathbb{Z}^2, 8, 4, (B \setminus X_b) \cup X_w)$ point p is simple, however, if the color of q is changed in the latter picture, then p is not a simple point in the new picture. As the simplicity is proved to be a local property by Theorem 1, we get that $q \in N_8^*(p)$, which means that p and q are 8-adjacent.   □

Theorem 6. *A general operator* $\mathcal{O}$ *is topology preserving in* $(8, 4)$ *pictures, if any subset of the points modified by* $\mathcal{O}$ *contained in a unit square is simple.*

Proof. The theorem is the easy consequence of Lemmas 1 and 2.   □

In Propositions 2-5, we formulate some further useful relationships which will be applied in the proof of our main theorem.

Proposition 2. *If* $\mathcal{O}$ *is a reduction that fulfills the conditions of Theorem 4 then any subset of the points modified by* $\mathcal{O}$ *is simple.*

Obviously, any subset of the points modified by $\mathcal{O}$ fulfills the conditions of Theorem 4. Hence, the proposition is satisfied by that theorem.

Proposition 3. *If* $\mathcal{O}$ *is an addition that satisfies the conditions of Theorem 5, then any subset of the points modified by* $\mathcal{O}$ *is simple.*

Any subset of the points modified by $\mathcal{O}$ fulfills the conditions of Theorem 5, Hence, the proposition holds by that theorem.

Proposition 4. *If sets* S *and* $Q \subset S$ *are simple sets, then* $S \setminus Q$ *is also a simple set.*

Indirectly, let us assume that $S \setminus Q$ is not simple. This means by the definition of simple sets that the modification of $S \setminus Q$ is not topology preserving. As the modification of S and Q is topology preserving, and the former one is equal with the sequential modification of Q and $S \setminus Q$, this can only be possible if changing $S \setminus Q$ also preserves the topology. However, this leads us to a contradiction.

Proposition 5. *If three points p, q, and r form a small triangle with corner point q in a picture, where p is a simple point and q is a black point, then p remains simple after r is modified.*

Point r is not 4-adjacent to p, i.e., $r \notin N_4^*(p)$, the color r is not important in order to decide whether p fulfills Condition 1 of Theorem 1 or not. Furthermore, q and r are 4-adjacent points, therefore the number of objects contained by $N_8^*(p)$, which is examined by Condition 2 of Theorem 1 is not influenced by the color of r. Hence, the proposition holds by Theorem 1.

In the proof of our main theorem we will use the notation $|Q|$ to refer to the count of elements in Q.

Theorem 7. *A general operator $\mathcal{O}$ is topology preserving in $(8,4)$ pictures, if it fulfills the following conditions:*

1. *Only simple points are modified by $\mathcal{O}$.*
2. *For any two 4-adjacent black points p and q deleted by $\mathcal{O}$, set $\{p,q\}$ is simple.*
3. *For any two 8-adjacent points $p \in \mathbb{Z}^2 \setminus B$ and $q \in \mathbb{Z}^2$ modified by $\mathcal{O}$, set $\{p,q\}$ is simple.*
4. *If $\mathcal{O}$ modifies a small triangle $T \not\subset B$ whose corner point is black, then T forms a simple set.*
5. *$\mathcal{O}$ does not delete completely any object contained in a unit square.*

Remark 1. Note that if $\mathcal{O}$ only modifies black points, i.e., it is a reduction, then Conditions 1, 2, and 5 imply the conditions of Theorem 4 , while if $\mathcal{O}$ only changes white points, i.e., it is an addition, then from Conditions 1 and 3 we get back the criteria of Theorem 5. Therefore, in these special cases the topological correctness of $\mathcal{O}$ directly follows from Theorem 4 or from Theorem 5 .

Now let us prove Theorem 7.

Proof. Let $\mathcal{P} = (\mathbb{Z}^2, 8, 4, B)$ be a picture, and let us denote S the set of points modified by $\mathcal{O}$ in $\mathcal{P}$. We refer by $\mathcal{A}$ to the addition that modifies the points of $S \cap (\mathbb{Z}^2 \setminus B)$ in picture $\mathcal{P}$, and we denote by $\mathcal{R}$ the reduction that deletes the points of $S \cap B$.

Let Q be a set of points modified by $\mathcal{O}$ that is contained in a unit square. By Theorem 6, it is sufficient to show that Q is a simple set.

If $|Q| = 1$, then by Condition 1, it is trivially simple. If Q contains only black points or only white points, then Q is a subset of the points modified by either $\mathcal{R}$ or $\mathcal{A}$. In this case, the simplicity of Q follows from Remark 1 and Propositions 2 and 3.

Let us suppose that Q contains at least one black point and at least one white point.

- If $|Q| = 2$, then Q is simple by Condition 3.
- If $|Q| = 3$, then Q is a small triangle. Let $Q = \{p, q, r\}$ with corner point q. If q is black, then Q is a simple set by Condition 4. Let us assume that q is white. By Condition 3 both p and r are simple in picture $(\mathbb{Z}^2, 8, 4, B \cup \{q\})$, and by Proposition 5, p is also simple in both of the pictures $(\mathbb{Z}^2, 8, 4, (B \setminus \{q\}) \cup \{r\})$ and $(\mathbb{Z}^2, 8, 4, (B \setminus \{q\}) \setminus \{r\})$. Hence Q is a simple set.

- Let us examine the case $|Q| = 4$. Let $Q = \{p, q, r, s\}$ such that q and s are both 4-adjacent to p. (This obviously also means that r is 8-adjacent but not 4-adjacent to p.) Without loss of generality, let us assume that q is white. Using the proof for the case $|Q| = 3$, we can show that the sets $\{p, q, s\}$ and $\{q, r, s\}$ are simple in $\mathcal{P}$, and by Proposition 4, both p and r are simple points in the picture that we get from $(\mathbb{Z}^2, 8, 4, B \cup \{q\})$ if we change the original color of s. Let us denote this latter picture $\mathcal{P}'$. Points p, q, and r in picture $\mathcal{P}'$ constitute a small triangle with black corner point q. Hence, by Proposition 5 point p is also simple in the picture that we get from $\mathcal{P}'$ by changing the color of r, which means that set Q is simple.

$\square$

4   An Example for a Topology Preserving Operator

Here we introduce a simple general operator working on (8,4) pictures, denoted by $\mathcal{T}$, and we prove its topological correctness based on the conditions of Theorem 7. $\mathcal{T}$ is defined with the four templates in Fig. 4. This means that $\mathcal{T}$ changes the color of a point p in an (8,4) picture if and only if the 3×3 neighborhood of p matches one of the given templates. (Note that the templates T_1 and T_2 are used for changing the color of some black points into white, while the remaining ones, T_3 and T_4 serve for turning some white points into black elements.)

In the later proof we will refer to the four elements in $N_4^*(p)$ by the notions $u(p)$, $l(p)$, $d(p)$, $r(p)$, see Fig. 5.

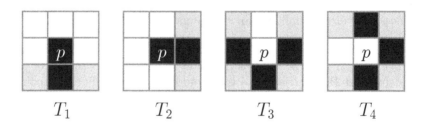

$$T_1 \qquad\qquad T_2 \qquad\qquad T_3 \qquad\qquad T_4$$

Fig. 4. Templates of the operator $\mathcal{T}$. Black and white positions of the templates refer to black and white points, respectively, while the cells depicted in gray represent points that can be either black or white.

Here we formulate some properties of the templates in Fig. 4, which will play a key role in the proof of the topological correctness of $\mathcal{T}$ in Theorem 8. All of them can be readily seen by the careful investigation of the templates.

Proposition 6. *If $\mathcal{T}$ modifies a black point p matched by template T_1, then $u(p)$, $u(r(p))$, $u(l(p))$, and $d(p)$ may not be matched by any of the templates of $\mathcal{T}$.*

Proposition 7. *If $\mathcal{T}$ modifies a black point p matched by template T_2, then $l(p)$, $l(u(p))$, $l(d(p))$, and $r(p)$ may not be matched by any of the templates of $\mathcal{T}$.*

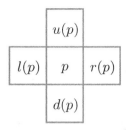

Fig. 5. Notions for the elements in $N_4^*(p)$

Proposition 8. *If a black point p is deleted by $\mathcal{T}$, then no one black point in $N_4^*(p)$ is deleted by $\mathcal{T}$.*

Theorem 8. *Operator $\mathcal{T}$ is topology preserving.*

Proof. It is sufficient to show that $\mathcal{T}$ satisfies the conditions of Theorem 7 for any picture $(\mathbb{Z}^2, 8, 4, B)$.

- It is easy to see by Theorem 1 that all templates of $\mathcal{T}$ represent simple points, therefore Condition 1 of Theorem 7 holds.
- If $\mathcal{T}$ removes a point p matched by template T_1, then by Proposition 6, $d(p)$ is not removed. Similarly, if p is matched by template T_2, $r(p)$ is not deleted by Proposition 7. Therefore, $\mathcal{T}$ does not delete any pair of 4-adjacent points, hence Condition 2 of Theorem 7.
- Let us suppose that $\mathcal{T}$ modifies a pair of 8-adjacent points $\{p, q\}$, where p is black and q is white. If p is matched by template T_1, then by Proposition 6, for any white point $q \in N_8^*(p)$ changed by $\mathcal{T}$, $q \in \{l(p), r(p), d(l(p)), d(r(p))\}$. As in template T_1, $d(p)$ is black, and $d(p)$ is 8-adjacent with q in all possible cases, therefore Condition 1 of Theorem 1 holds for p in picture $(\mathbb{Z}^2, 8, 4, B \cup \{q\})$, and as $N_4^*(p)$ contains more than one white point, p also fulfills Condition 2 of Theorem 1 in the latter picture. Thus, p remains simple by Theorem 1 after the modification of q, which means that set $\{p, q\}$ is simple. Using Proposition 7, we can show the former property similarly if p is matched by template T_2. This concludes Condition 3 of Theorem 7.
- We have already seen that $\mathcal{T}$ does not remove a 4-adjacent pair of black points. This property combined with Propositions 6 and 7 implies that $\mathcal{T}$ does not change any small triangle with a black corner point, hence Condition 4 of Theorem 7 trivially holds.
- By Proposition 8 we can state that if a black point of an object is deleted by $\mathcal{T}$, then there is exactly one black point in $N_4^*(p)$ that is preserved, which concludes Condition 5 of Theorem 7.

□

Let us perform operator $\mathcal{T}$ on the small binary image in Fig. 6(a) of an object containing a cavity. Studying the result of the transformation in Fig. 6(b), we can notice a nice smoothing effect of $\mathcal{T}$ on the upper and the left sides of object contour.

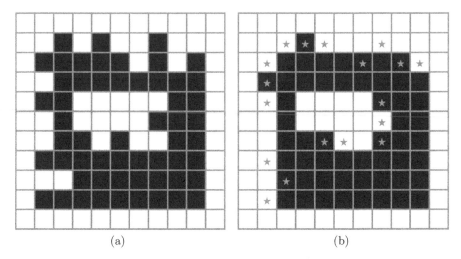

Fig. 6. The original picture (a) and the result produced by $\mathcal{T}$ (b). Modified points are denoted with stars.

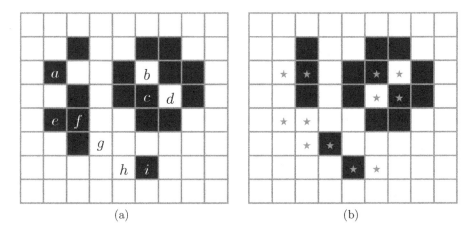

Fig. 7. The original picture (a) and the result produced by a topology preserving operator $\mathcal{U}$ (b). Modified points are marked stars. Points denoted a, b, c, and d are not simple, i.e., Condition 1 of Theorem 7 is not satisfied. Set $\{e, f\}$ does not fulfill Condition 2, while sets $\{b, c\}$, $\{c, d\}$, $\{b, d\}$, and $\{g, h\}$ violate Condition 3. Furthermore Condition 4 is not fulfilled for the set $\{b, c, d\}$, while Condition 5 does not hold for the isolated object point i.

It is important to emphasize that the conditions of Theorem 7 are only sufficient but not necessary. Let us consider for example an operator $\mathcal{U}$ that turns the small (8,4) picture in Fig. 7(a) into the image in Fig. 7(b). It is obvious that, despite that none of the conditions of Theorem 7 holds, $\mathcal{U}$ is topology preserving.

5   Conclusion and Future Work

In this work we introduced some sufficient conditions which can be applied to verify the topological correctness of general 2D operators for (8,4) pictures. These criteria can be regarded as the very first extensions of the ones for reductions presented in [14] and for additions proposed in [8].

As the next step of our research we intend to formulate similar conditions for general 3D operators working on (26,6) pictures. Another interesting problem we plan to investigate is the necessity of topology preservation for reductions, additions, and general operators. Simplicity of the unique points is an important property if modifications along the object contour are only allowed, which, for instance, is the case for thinning algorithms. However, generally spoken it is neither a necessary nor a sufficient condition in the view of topology preservation. Therefore, we consider it expedient to divide general operators into two groups according to whether they only change simple points or not and to examine necessity of topological correcntess for that two kinds of operators separately.

Later, we also would like to apply those general conditions for designing a new contour smoothing algorithm for 3D binary objects, which not only removes small protrusions but also fills small bays. Furthermore, using such an algorithm, it is possible to give an improved version of the scheme for thinning combined with iteration-level contour smoothing as described in [12].

Acknowledgments. This research was supported by the European Union and the State of Hungary, co-financed by the European Social Fund in the framework of TÁMOP 4.2.4. A/2-11-1-2012-0001 "National Excellence Program".

References

1. Hall, R.W.: Parallel connectivity–preserving thinning algorithm. In: Kong, T.Y., Rosenfeld, A. (eds.) Topological Algorithms for Digital Image Processing, Machine Intelligence and Pattern Recognition, vol. 19, pp. 145–179. Elsevier Science (1996)
2. Hall, R.W., Kong, T.Y., Rosenfeld, A.: Shrinking binary images. In: Kong, T.Y., Rosenfeld, A. (eds.) Topological Algorithms for Digital Image Processing, Machine Intelligence and Pattern Recognition, vol. 19, pp. 31–98. Elsevier Science (1996)
3. Herman, G.T.: Geometry of digital spaces. Birkhäuser, Boston (1998)
4. Kong, T.Y.: Minimal non-simple and minimal non-cosimple sets in binary images on cell complexes. In: Kuba, A., Nyúl, L.G., Palágyi, K. (eds.) DGCI 2006. LNCS, vol. 4245, pp. 169–188. Springer, Heidelberg (2006)
5. Kong, T.Y.: On topology preservation in 2-d and 3-d thinning. International Journal of Pattern Recognition and Artificial Intelligence 9, 813–844 (1995)
6. Kong, T.Y., Rosenfeld, A.: Digital topology: Introduction and survey. Computer Vision, Graphics, and Image Processing 48, 357–393 (1989)
7. Lam, L., Lee, S.-W., Suen, C.Y.: Thinning methodologies – A comprehensive survey. IEEE Trans. Pattern Analysis and Machine Intelligence 14, 869–885 (1992)
8. Kardos, P., Palágyi, K.: Sufficient conditions for topology preserving additions and general operators. In: Proc. 14th Int. Conf. Computer Graphics and Imaging, CGIM 2013, pp. 107–114. IASTED ACTA Press (2013)

9. Ma, C.M.: On topology preservation in 3D thinning. CVGIP: Image Understanding 59, 328–339 (1994)
10. Marchand-Maillet, S., Sharaiha, Y.M.: Binary digital image processing – A discrete approach. Academic Press (2000)
11. Németh, G., Kardos, P., Palágyi, K.: 2D parallel thinning and shrinking based on sufficient conditions for topology preservation. Acta Cybernetica 20, 125–144 (2011)
12. Németh, G., Kardos, P., Palágyi, K.: Thinning combined with iteration-by-iteration smoothing for 3D binary images. Graphical Models 73, 335–345 (2011)
13. Palágyi, K., Németh, G., Kardos, P.: Topology Preserving Parallel 3D Thinning Algorithms. In: Brimkov, V.E., Barneva, R.P. (eds.) Digital Geometry Algorithms. Theoretical Foundations and Applications to Computational Imaging, pp. 165–188. Springer (2012)
14. Ronse, C.: Minimal test patterns for connectivity preservation in parallel thinning algorithms for binary digital images. Discrete Applied Mathematics 21, 67–79 (1988)
15. Rosenfeld, A.: Arcs and curves in digital pictures. Journal of the ACM 20(1), 81–87 (1973)
16. Rosenfeld, A.: Digital topology. The American Mathematical Monthly 86(8), 621–630 (1979)
17. Serra, J.: Image Analysis and Mathematical Morphology. Academic Press (1982)
18. Suen, C.Y., Wang, P.S.P.: Thinning Methodologies for Pattern Recognition. Series in Machine Perception and Artificial Intelligence, vol. 8. World Scientific, Singapore (1994)

Speed Comparison
of Segmentation Evaluation Methods

Stepan Srubar

VŠB-Technical University of Ostrava, Czech Republic
`stepan.srubar@vsb.cz`

Abstract. Segmentation algorithms are widely used in image processing and there is a definite need for good quality segmentation algorithms. In order to assess which segmentation algorithms are good for our tasks, we need to measure their quality. This is done by evaluation methods. Still, we have the same problem. There are several evaluation methods, but which are good and fast enough? This article measures the quality and speed of some evaluation methods and shows that there are large differences between them.

Keywords: Segmentation, Evaluation, Quality, Speed.

1 Introduction

Segmentation is an important part of image processing. It separates the image into segments. Each segment can be measured for some specific properties like perimeter, area, curvature or it can be used for processing or modifying image information. For best results, we need high quality segmentation produced by a segmentation algorithm. The higher the quality of the segmentation algorithm, the higher the quality of the segmentation itself. In order to measure the quality of the algorithm various evaluation methods are applied.

The evaluation method must not only give reliable results, but to produce them fast, in seconds or even milliseconds. The latter requirement is due to the fact that the measuring of the quality may need more than one evaluation. Moreover, segmentation algorithms use parameters. Hence, we may want to know which of the thousands combinations of parameters are the best ones. In such a case, the speed of the evaluation method becomes critical.

There could be large differences in the quality and speed of evaluation methods. This article will show how to measure the quality and speed and will present the comparative results of some recent and older methods. The next section describes the evaluation methods used for final measurement. The third section shows the data set and describes the measuring of quality and speed. The description of results and their representation can be found in the fourth section.

2 Evaluation Methods

Before we start with the description of methods, we will introduce some basic definitions to make the paper self-contained. Segmentation is groupping of image

R.P. Barneva, V.E. Brimkov, and J. Šlapal (Eds.): IWCIA 2014, LNCS 8466, pp. 113–122, 2014.
© Springer International Publishing Switzerland 2014

pixels into segments. Therefore, each pixel belongs to one segment. In other words, the segments are defined by their pixels. Thus, similarly to the image, we can decide whether a pixel belongs to a segment or not. We can also measure the distance of pixels in the segmentation and segments's sizes. The evaluation method is a function, which takes two segmentations and outputs a number. This number expresses the similarity of these two segmentations.

A pixel may belong to one segment in the first segmentation and to a different segment in the second segmentation. We call these segments correspondent because they have something in common. Note that the correspondent segments are not from the same segmentation. The correspondence is often defined by the evaluation method according to specific properties.

Each method takes two segmentations for comparison and computes the result. However, the result can be different if these two segmentations are swapped. In such a case we call the method asymmetric. If the results are not dependent on the order, we call it symmetric.

This article includes 30 evaluation methods. For clarity, each method is denoted by an abbreviation. Moreover, these methods are grouped into six categories, which are provided in the following subsections. This categorization is based on an approach or on the properties of the methods such as the usage of a segment and intersection sizes, distances of borders and segments, statistical measurement, and the number of segments.

2.1   Segment and Intersection Size Based Methods

The difference or ratio of the size of a segment and its intersection with another segment is a common way of evaluation the rate of correspondence of segments. This approach was used in methods SM_1, SM_2 [24], LCE, GCE, BCE [7,6], $GBCE$ [21], HD [3], PD [15], L [5], VD [20], and MH [9]. Despite the naming Hamming Distance (HD) and Partition Distance (PD), both methods use only the segments' size and their intersections. The method VD is a metric.

A little bit different approach was chosen by Pal and Bhandari in [12]. Their Symmetric Divergence SYD uses natural logarithm. Therefore, the method is not symmetric, eventhough one could assume so from its name.

2.2   Methods Using Distance Measure

One of the easiest ways to include segments' shape is to compute the distance between pixels. The most used distance is the distance of a pixel to the nearest pixel of the corresponding segment. It is used in methods YD [24], F [16], MC [11], and SD_1, SD_2 [21]. All the pixels from the intersection of the corresponding segments have zero distance, therefore, they do not influence the result.

Different type of distance is the distance of border pixels. Border pixels are pixels that belong on the border of the segment. Searching for these pixels increases complexity of implementation and decreases the speed. Such methods are $SFOM$ [13], and SD_3, SD_4 [21]

2.3   Methods Based on Counting of Couples

A randomly placed couple of points in a segmentation leads to only two possibilities. Either they lay in the same segment or in different segments. In the case of two segmentations, we get four possibilities shown in the Figure 1. We can count the number of couples of each type separately, so we get four numbers according to the Figure 1: N_{11}, N_{10}, N_{01}, and N_{00}. The following methods use only these statistical numbers.

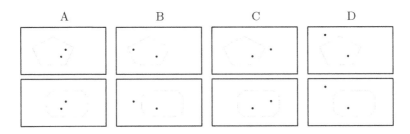

Fig. 1. All four possible types of positions (A-D) of two points in two different segmentations

In the case of comparison of two identical segmentations, both N_{10} and N_{01} are zero. This property is base of Mirkin metric [10]. When these two numbers are zero then the sum of others must be equal to number of all couples. This is approach of Rand index [14], later extended [18,19] for fuzzy segmentations (PRI).

The following methods use combination of three numbers: JC [1], FM [2], W [22]. None of the stated methods use all four numbers.

2.4   Methods with Statistical Approach

The previous subsection includes also methods with statistical approach but in a specific way. The following methods are built on different bases.

Yasnoff and Bacus proposed Object Count Agreement (OCA) in [23]. They use gamma function and integrals and, therefore, the complexity of computation is high. They suppose grouping of segments into classes so that each class can represent some discontinuous object. Still, we don't have segmentations with discontinuous objects, so each class will have only one segment.

Not only the entropy of a data set, but also the entropy of segmentations can be computed. Normalization of the latter will bring the Normalized Mutual Information (NMI) [17]. Further improvement can lead to metric like in the case of Variation of Information (VI) [8].

2.5 Graph Theory Methods

Segments can be seen as nodes of a graph. Edges connect segments and the weight of an edge is the intersection size. Because segments in segmentation do not overlap, edges are between segments from the opposite segmentation only. Now we need to find correspondent segments. This is represented by one edge per node. So we need to remove some edges from the graph. Jiang et al. in [4] propose to find maximum-weight bipartite graph. That will keep one edge per node at most and maximize the total sum of intersection sizes.

2.6 Number of Segments Methods

Methods computing the number of segments are easy to implement and also very fast. One of them is the Fragmentation proposed in [16]

$$FRAG(S_1, S_2, p, q) = \frac{1}{1 + p \cdot ||S_1| - |S_2||^q},$$ (1)

where $p > 0$ and $q > 0$ are scaling parameters and $|S_1|$ and $|S_2|$ are number of segments of segmentations S_1 and S_2, respectively.

3 Methodology of Comparison

All implemented methods were evaluated on an image data set [7]. For speed measurement, the number of segmentation evaluations done in the first minute was counted. For quality measurement, the whole database was used. It consists of sample images (481 x 321 pixels) and their ground truth segmentations (see Figure 2). Each image has four ground truth segmentations at least. Therefore,

Fig. 2. Example of an image and its ground truth segmentations from Berkeley segmentation database

we could compare segmentations which belong to the same image. This is a crucial property for the comparison.

Images in the database are longitudinal and perpendicular. Perpendicular segmentations were rotated to be comparable with longitudinal ones. Symmetric methods have twice less results than asymmetric methods, but the resulting error rates are not dependent on the number of comparisons; thus the results from these two groups were merged.

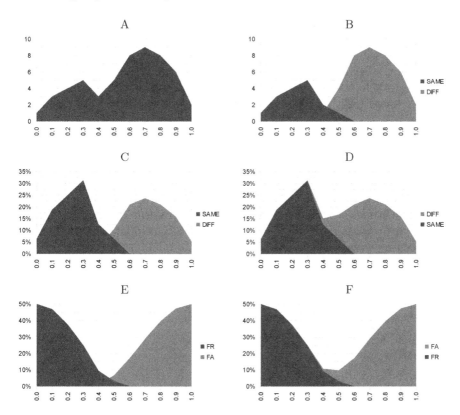

Fig. 3. Processing data on an example. A) histogram of results; B) separation of results according to same/different segmentations; C) normalized results; D) normalized results stacked (summed); E) false acceptance and rejection rates; F) final representation - stacked false rates.

The comparison of two segmentations results in a number. After all comparisons, we have a set of numbers that can be represented in a histogram like the one in Figure 3A. Since we know which results come from comparing of two segmentations of the same image or of different images, we can divide the histogram into two histograms (Figure 3B). There were much more comparisons of segmentations of different images so we normalized these results (Figure 3C). We could represent the same data in a stacked graph which is just a sum of values (Figure 3D).

It seems natural to set the threshold to 0.4 because this is the lowest point of the graph. This threshold would divide the results of the same and of different images. Such shortcut is misleading because we want to minimize the number of results on the wrong side of the threshold. Therefore, we need to do integration of the previous graph (Figure 3E). It represents the error rates according to the threshold. The results are false acceptance (FA) and rejection (FR) rates. False acceptance is an error rate of the method which represents the portion of results of segmentations of different images with very low numbers as the output of the method. In other words, the results should be higher when the segmentations do not belong to the same image.

Now we can see that the lowest point moved between 0.4 and 0.5. Such point means equality of error rates. However, we need the lowest possible error rate in total, thus, we find the sum of these results in the stacked graph in Figure 3F. Finally, we found that the optimal threshold would be 0.5 with total error rate just below 10%.

The final result is sum of both error rates at point 0.5 which represents the best possible threshold for results of this sample evaluation method. The better the method, the better the separation of results, and the lower the lowest error rate. This processing of results was done for all methods and only the lowest error rates are presented. Such error rates represent the quality of the methods.

Speed of methods (comparisons per second) was measured on Athlon QL-60 dual core processor. Algorithms were compiled with OpenMP support thus the measurement run on both cores simultaneously. To minimize the time spent on accessing data, all the segmentations were saved in RAM disk. Still, the processor loading caused by RAM disk operations could reach up to twenty percent when measuring the fastest methods. Therefore, the fastest methods could be even faster.

4   Results

The quality of methods differs significantly. The results are presented in Figure 4. The worst score of a method is 50% because it equals guessing. The best method is SD_3. According to the year of publication, we could show the quality of methods on time axis (Figure 5). As one can see only four methods were able to beat historically the best methods of that time: PRI (Rand index, 1971), FM (1983), VD (1995), SD_3 (2011).

A high quality method could be practically unusable if it is very slow. Therefore, the speed of all methods was measured as it can be seen in the Figure 6. (The results are sorted by speed.) If we want to pick up a method, we would select a method which is fast enough and of a high quality. In other words, we would pick a method with a minimal speed which is enough for our needs and then we would look for a method which is fast and with the best quality. This condition is fulfilled by only four methods: SD_3, VD, MH and $FRAG$. They are shown in different colors in the graph. Each of these four methods has better quality than all other fast methods. It is interesting to note the speed difference

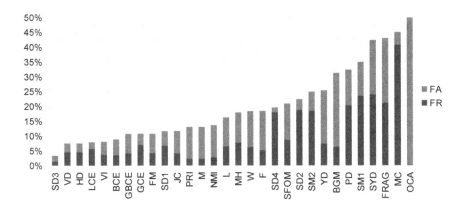

Fig. 4. All methods sorted according to their quality. The lowest rates of false acceptance (FA) and false rejection (FR) are showed.

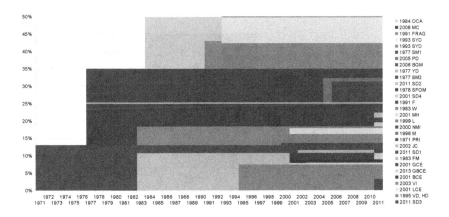

Fig. 5. History of quality of the methods. Each rectangle represents a method by the year of its publication.

between SD_3 and VD. As they are the best methods, there could be no other better method between them. We could also see that the fastest methods need not to be the worst. For example, the third fastest method is the second best in quality.

In the same way as the history of quality was presented, the history of speed is shown in the Figure 7. Historically, the first method (Rand index) was quite fast, thus, it took nearly 25 years to discover a faster method with usable quality (HD).

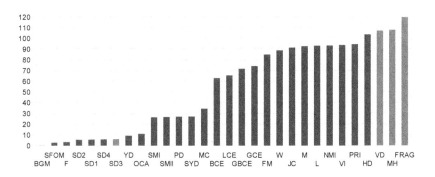

Fig. 6. All methods sorted according to their speed (comparisons per second). Four methods are shown in different color because they bring higher quality in comparison to all faster methods.

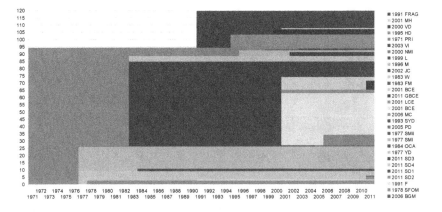

Fig. 7. History of speed of methods. Each rectangle represents a method by the year of its publication.

5 Conclusion

A lot of segmentation evaluation methods exist, but some of them are unusable due to poor quality and some of them—due to very low speed. All the presented methods were described and measured for their quality and speed. The final comparison is quite surprising, because the majority of methods are both slower and give worse results then some other methods. All in all, just two methods can be recommended: SD_3 and VD. The first one has the best quality with acceptable speed while the latter one brings very high speed at cost of little drop in quality.

References

1. Ben-Hur, A., Elisseeff, A., Guyon, I.: A stability based method for discovering structure in clustered data. In: Pacific Symposium on Biocomputing, pp. 6–17 (2002), http://view.ncbi.nlm.nih.gov/pubmed/11928511
2. Fowlkes, E.B., Mallows, C.L.: A Method for Comparing Two Hierarchical Clusterings. Journal of the American Statistical Association 78(383), 553–569 (1983), http://dx.doi.org/10.2307/2288117
3. Huang, Q., Dom, B.: Quantitative methods of evaluating image segmentation. In: Proceedings of the International Conference on Image Processing, vol. 3, pp. 53–56 (1995), http://dx.doi.org/10.1109/ICIP.1995.537578
4. Jiang, X., Marti, C., Irniger, C., Bunke, H.: Distance measures for image segmentation evaluation. EURASIP J. Appl. Signal Process. 2006, 209 (2006), http://dx.doi.org/10.1155/ASP/2006/35909
5. Larsen, B., Aone, C.: Fast and effective text mining using linear-time document clustering. In: Proceedings of the Fifth ACM SIGKDD International Conference on Knowledge Discovery and Data Mining, KDD 1999, pp. 16–22. ACM, New York (1999), http://doi.acm.org/10.1145/312129.312186
6. Martin, D.R.: An empirical approach to grouping and segmentation. Phd dissertation, University of California, Berkeley (June-August 2002)
7. Martin, D.R., Fowlkes, C., Tal, D., Malik, J.: A database of human segmented natural images and its application to evaluating segmentation algorithms and measuring ecological statistics. Tech. Rep. UCB/CSD-01-1133, EECS Department, University of California, Berkeley (January 2001), http://www.eecs.berkeley.edu/Pubs/TechRpts/2001/6434.html
8. Meilă, M.: Comparing Clusterings by the Variation of Information. In: Schölkopf, B., Warmuth, M.K. (eds.) COLT/Kernel 2003. LNCS (LNAI), vol. 2777, pp. 173–187. Springer, Heidelberg (2003)
9. Meilă, M., Heckerman, D.: An experimental comparison of model-based clustering methods. Mach. Learn. 42, 9–29 (2001), http://portal.acm.org/citation.cfm?id=599609.599627
10. Mirkin, B.G.: Mathematical Classification and Clustering. Kluwer Academic Publishers, Dordrecht (1996)
11. Monteiro, F.C., Campilho, A.C.: Performance evaluation of image segmentation. In: Campilho, A., Kamel, M. (eds.) ICIAR 2006. LNCS, vol. 4141, pp. 248–259. Springer, Heidelberg (2006), http://dx.doi.org/10.1007/11867586_24
12. Pal, N.R., Bhandari, D.: Image thresholding: some new techniques. Signal Process. 33, 139–158 (1993), http://dx.doi.org/10.1016/0165-1684{93}90107-L

13. Pratt, W.K.: Digital Image Processing. John Wiley & Sons, Inc., New York (1978)
14. Rand, W.M.: Objective Criteria for the Evaluation of Clustering Methods. Journal of the American Statistical Association 66(336), 846–850 (1971), http://dx.doi.org/10.2307/2284239
15. dos Santos Cardoso, J., Corte-Real, L.: Toward a generic evaluation of image segmentation. IEEE Transactions on Image Processing 14(11), 1773–1782 (2005), http://dx.doi.org/10.1109/TIP.2005.854491
16. Strasters, K.C., Gerbrands, J.J.: Three-dimensional image segmentation using a split, merge and group approach. Pattern Recogn. Lett. 12, 307–325 (1991), http://dx.doi.org/10.1016/0167-8655(91)90414-H
17. Strehl, A., Ghosh, J., Mooney, R.: Impact of Similarity Measures on Web-page Clustering. In: Proceedings of the 17th National Conference on Artificial Intelligence: Workshop of Artificial Intelligence for Web Search (AAAI 2000), Austin, Texas, USA, July 30-31, pp. 58–64. AAAI (July 2000)
18. Unnikrishnan, R., Pantofaru, C., Hebert, M.: A measure for objective evaluation of image segmentation algorithms. In: Proceedings of the 2005 IEEE Computer Society Conference on Computer Vision and Pattern Recognition (CVPR 2005) - Workshops, vol. 03, p. 34+. IEEE Computer Society, Washington, DC (2005) http://portal.acm.org/citation.cfm?id=1099539.1099935
19. Unnikrishnan, R., Pantofaru, C., Hebert, M.: Toward Objective Evaluation of Image Segmentation Algorithms. IEEE Transactions on Pattern Analysis and Machine Intelligence 29(6), 929–944 (2007), http://dx.doi.org/10.1109/TPAMI.2007.1046
20. Van Dongen, S.: Performance criteria for graph clustering and markov cluster experiments. Tech. rep., National Research Institute for Mathematics and Computer Science, Amsterdam, Netherlands (2000)
21. Šrubař, V.: Quality measurement of image segmentation evaluation methods. In: 8th International Conference on Signal Image Technology and Internet Based Systems, SITIS 2012r, pp. 254–258 (2012), http://www.scopus.com
22. Wallace, D.L.: A Method for Comparing Two Hierarchical Clusterings: Comment. Journal of the American Statistical Association 78(383), 569–576 (1983)
23. Yasnoff, W.A., Bacus, J.W.: Scene segmentation algorithm development using error measures. AOCH 9, 45–58 (1984)
24. Yasnoff, W.A., Mui, J.K., Bacus, J.W.: Error measures for scene segmentation. Pattern Recognition 9(4), 217–231 (1977), http://www.sciencedirect.com/science/article/B6V14-48MPGVT-27/2/b7756a79dfd9490c6e74feb9b6f77941

A Variant of Pure Two-Dimensional Context-Free Grammars Generating Picture Languages

Zbyněk Křivka[1], Carlos Martín-Vide[2],
Alexander Meduna[1], and K.G. Subramanian[3]

[1] IT4Innovations Centre of Excellence, Faculty of Information Technology,
Brno University of Technology,
Božetěchova 1/2, 612 66 Brno, Czech Republic
[2] Research Group on Mathematical Linguistics, Rovira i Virgili University,
Av Catalunya, 35, 43002 Tarragona, Spain
[3] School of Computer Sciences, Universiti Sains Malaysia,
11800 Penang, Malaysia

Abstract. Considering a large variety of approaches in generating picture languages, the notion of pure two-dimensional context-free grammar ($P2DCFG$) represents a simple yet expressive non-isometric language generator of picture arrays. In the present paper, we introduce a new variant of $P2DCFG$s that generates picture arrays in a leftmost way. We concentrate our attention on determining their generative power by comparing it with the power of other picture generators. We also examine the power of these generators that regulate rewriting by control languages.

Keywords: Two-dimensional arrays, Array grammars, Pure grammars, Context-free grammars.

1 Introduction

Recently, several two-dimensional (2D) picture generating grammars [4,10,11,16,17] have been introduced and investigated. The introduction of these grammars has been motivated by problem areas ranging from tiling patterns through certain floor designs up to geometric shapes. These 2D grammars have been mainly developed based on the concepts and techniques of string grammar theory. In essence, there exist two basic variants–(i) isometric array grammars in which geometric shape of the rewritten portion of the array is preserved, and (ii) non-isometric array grammars that can alter the geometric shape. In the present paper, we discuss *pure 2D context-free grammar* ($P2DCFG$), which is related to (ii) (see [15]). In essence, the notion of $P2DCFG$ involves only terminal symbols as in any pure grammar [5] and tables of context-free (CF) rules. In this grammar, all the symbols in a column or a row of a rectangular picture array are rewritten by CF rules with all symbols being replaced in parallel by

R.P. Barneva, V.E. Brimkov, and J. Šlapal (Eds.): IWCIA 2014, LNCS 8466, pp. 123–133, 2014.

strings of equal length, thus maintaining the rectangular form of the array. In [1,2,14], various properties of this 2D grammar model are studied.

In string grammars, leftmost derivations (see, for example, [3,6,8]) have been extensively studied. Recall that in the case of context-free grammars, corresponding to an ordinary derivation, there is an equivalent leftmost derivation that rewrites only the leftmost nonterminal in a sentential form (see [7]). In this paper, we discuss leftmost rewriting in terms of $P2DCFG$. In other words, while a $P2DCFG$ allows rewriting any column or any row of a picture array by the rules of an applicable column rule table or row rule table respectively, in the variant under the investigation in the present paper, only the leftmost column or the uppermost row of an array is rewritten. We refer to the $P2DCFG$ working under this derivation mode as $(l/u)P2DCFG$ and the corresponding family of picture languages generated by them as $(l/u)P2DCFL$. We demonstrate that $(l/u)P2DCFL$ and the family of picture languages generated by $P2DCFG$s are incomparable, and that $(l/u)P2DCFL$ is not closed under union and intersection. The effect of regulated rewriting in $(l/u)P2DCFG$s by control languages is also examined, and it is demonstrated that this regulation results into an increase in the generative power.

2 Preliminaries

For notions related to formal language theory we refer to [7,12,13] and for array grammars and two-dimensional languages we refer to [4].

A word or a string $w = a_1 a_2 \ldots a_n$ $(n \geq 1)$ over a finite alphabet Σ is a sequence of symbols from Σ. The length of a word w is denoted by $|w|$. The set of all words over Σ, including the empty word λ with no symbols, is denoted by Σ^*. For any word $w = a_1 a_2 \ldots a_n$, we denote by ${}^t w$ the word w written vertically, with ^{t} having lower precedence than concatenation, so that ${}^t w = {}^t(w)$.

For example, if $w = abb$ over $\{a, b\}$, then ${}^t w$ is $\begin{matrix} a \\ b \\ b \end{matrix}$. A two-dimensional array (also called picture array or picture) is a rectangular $m \times n$ array p over Σ of the form

$$p = \begin{matrix} p(1,1) & \cdots & p(1,n) \\ \vdots & \ddots & \vdots \\ p(m,1) & \cdots & p(m,n) \end{matrix}$$

where each $p(i,j) \in \Sigma, 1 \leq i \leq m, 1 \leq j \leq n$. A pixel is an element $p(i,j)$ of p. $|p|_{row}$ and $|p|_{col}$ denote the number of rows of p and the number of columns of p, respectively. The size of p is the pair $(|p|_{row}, |p|_{col})$. The set of all rectangular arrays over Σ is denoted by Σ^{**}, which includes the empty array λ. $\Sigma^{++} = \Sigma^{**} - \{\lambda\}$. A picture language is a subset of Σ^{**}.

We now recall a pure 2D context-free grammar introduced in [14,15].

Definition 1. *A pure 2D context-free grammar (P2DCFG) is a 4-tuple*

$$G = (\Sigma, P_1, P_2, \mathcal{M}_0)$$

where

i) Σ is a finite alphabet of symbols;

ii) $P_1 = \{c_i \mid 1 \leq i \leq s_c\}$, where c_i is called a *column rule table* and s_c is some positive integer; each c_i is a finite set of context-free rules of the form $a \rightarrow \alpha, a \in \Sigma, \alpha \in \Sigma^*$ such that for any two rules $a \rightarrow \alpha, b \rightarrow \beta$ in c_i, we have $|\alpha| = |\beta|$ i.e. α and β have equal length;

iii) $P_2 = \{r_j \mid 1 \leq j \leq s_r\}$, where r_j, is called a *row rule table* and s_r is some positive integer; each r_j is a finite set of rules of the form $c \rightarrow {}^t\gamma, c \in \Sigma, \gamma \in \Sigma^*$ such that for any two rules $c \rightarrow {}^t\gamma, d \rightarrow {}^t\delta$ in r_j, we have $|\gamma| = |\delta|$;

iv) $\mathcal{M}_0 \subseteq \Sigma^{**} - \{\lambda\}$ is a finite set of axiom arrays.

A derivation in a *P2DCFG G* is defined as follows: Let $p, q \in \Sigma^{**}$. The picture q is derived from picture p in G, denoted by $p \Rightarrow q$, if q is obtained from p either *i*) by rewriting in parallel all the symbols in a column of p, each symbol by a rule in some column rule table or *ii*) rewriting in parallel all the symbols in a row of p, each symbol by a rule in some row rule table. All the rules used to rewrite a column (or row) have to belong to the same table.

The picture language generated by G is the set of picture arrays $L(G) = \{M \in \Sigma^{**} \mid M_0 \Rightarrow^* M$ for some $M_0 \in \mathcal{M}_0\}$. The family of picture languages generated by *P2DCFGs* is denoted by *P2DCFL*.

Example 1. Consider the *P2DCFG* $G_1 = (\Sigma, P_1, P_2, \{M_0\})$ where $\Sigma = \{a, b, e\}$, $P_1 = \{c\}, P_2 = \{r\}$, where $c = \{a \rightarrow bab, e \rightarrow aea\}$, $r = \left\{e \rightarrow \begin{smallmatrix} e \\ a \end{smallmatrix}, a \rightarrow \begin{smallmatrix} a \\ b \end{smallmatrix}\right\}$, and

$M_0 = \begin{smallmatrix} a\ e\ a \\ b\ a\ b \end{smallmatrix}$.

G_1 generates a picture language L_1 consisting of picture arrays p of size $(m, 2n + 1)$, $m \geq 2$, $n \geq 1$ with $p(1, j) = p(1, j + n + 1) = a$, for $1 \leq j \leq n$; $p(1, n + 1) = e$; $p(i, n + 1) = a$, for $2 \leq i \leq n$; $p(i, j) = b$, otherwise. A member of L_1 is shown in Figure 1.

$$
\begin{array}{ccccccc}
a & a & a & e & a & a & a \\
b & b & b & a & b & b & b \\
b & b & b & a & b & b & b \\
b & b & b & a & b & b & b \\
b & b & b & a & b & b & b
\end{array}
$$

Fig. 1. A picture in the language L_1

We note that the rows in the generated picture arrays of L_1 do not maintain any proportion to the columns since the application of the column rule table c can take place independent of the row rule table r. But the picture array will have an equal number of columns to the left and right of the middle column ${}^t(ea \ldots a)$.

We now recall a *P2DCFG* with a control language on the labels of the column rule and row rule tables in the *P2DCFG*, which is introduced in [14,15].

A *P2DCFG* with a regular control is $G^c = (G, \Gamma, \mathcal{C})$ where $G = (\Sigma, P_1, P_2, \mathcal{M}_0)$ is a *P2DCFG*, Γ is a set of labels of the tables of G, given by $\Gamma = P_1 \cup P_2$ and $\mathcal{C} \subseteq \Gamma^*$ is a regular (string) language. The words in Γ^* are called control words of G. Derivations $M_1 \Rightarrow_w M_2$ in G^c are done as in G except that if $w \in \Gamma^*$ and $w = l_1 l_2 \ldots l_m$, then the tables of rules with labels $l_1, l_2, \ldots$, and l_m are successively applied starting with the picture array M_1 to finally yield the picture array M_2. The picture array language generated by G^c consists of all picture arrays obtained from axiom arrays of G with the derivations controlled as described above. We denote the family of picture languages generated by *P2DCFGs* with regular control by $(R)P2DCFL$.

3 Pure 2D Context-Free Grammar with (l/u) Mode of Derivations

We now consider a variant in the rewriting process of a *P2DCFG*. The concept of leftmost derivation in a context-free grammar in string language theory, is well-known [12,13], especially in the context of LL parsers. In fact, in [9], the leftmost derivation concept is generalized to obtain derivation trees for context-sensitive grammars. On the other hand, leftmost derivations have been considered in other string grammars as well. For example, Meduna and Zemek [8] have studied the generative power of one-sided random context grammars working in the leftmost way. These studies, especially the study in [8], motivate to consider a corresponding notion of "leftmost kind" of derivation in pure 2D context-free grammars with a view to compare the resulting picture generative power with the *P2DCFG* [14] as well as to examine other kinds of results such as closure properties. The idea is to rewrite the leftmost column of a picture array by a column rule table or the uppermost row by a row rule table unlike the unrestricted way of rewriting any column or any row (if a column rule or row rule table is applicable) in a *P2DCFG*. This kind of a restriction on rewriting results in a picture language family which neither contains nor is contained in *P2DCFL*.

Definition 2. *Let $G = (\Sigma, P_1, P_2, \mathcal{M}_0)$ be a P2DCFG with the components as in Definition 1. An (l/u) mode of derivation of a picture array M_2 from M_1 in G, denoted by $\Rightarrow_{(l/u)}$, is a derivation in G such that only the leftmost column or the uppermost row of M_1 is rewritten using respectively, the column rule tables or the row rule tables, to yield M_2. The generated picture language is defined as in the case of a P2DCFG but with $\Rightarrow_{(l/u)}$ derivations. The family of picture languages generated by P2DCFGs under $\Rightarrow_{(l/u)}$ derivations is denoted by $(l/u)P2DCFL$. For convenience, we write $(l/u)P2DCFG$ to refer to P2DCFG with $\Rightarrow_{(l/u)}$ derivations.*

We illustrate with an example.

Example 2. Consider an $(l/u)P2DCFG$ $G_2 = (\Sigma, P_1, P_2, \{M_0\})$ where $\Sigma = \{a, b\}$, $P_1 = \{c\}$, $P_2 = \{r\}$ with $c = \{a \rightarrow ab, b \rightarrow ba\}$, $r = \left\{a \rightarrow \dfrac{a}{b}, b \rightarrow \dfrac{b}{a}\right\}$, and $M_0 = \dfrac{b\ a}{a\ b}$.

G_2 generates a picture language L_2 consisting of arrays p of size (m, n), $m \geq 2$, $n \geq 2$ with $p(1, 1) = b$; $p(1, j) = a$, for $2 \leq j \leq n$; $p(i, 1) = a$, for $2 \leq i \leq m$; $p(i, j) = b$, otherwise. A member of L_2 is shown in Figure 2. A sample derivation

$$
\begin{array}{llllll}
b & a & a & a & a & a \\
a & b & b & b & b & b \\
a & b & b & b & b & b \\
a & b & b & b & b & b \\
a & b & b & b & b & b
\end{array}
$$

Fig. 2. A picture array in the language L_2

in $(l/u)P2DCFG$ G_2 starting from M_0 and using the tables c, r, c, c in this order is shown in Figure 3. We note that in this derivation (unlike in a derivation in a $P2DCFG$), the application of the column rule table c rewrites all symbols in the leftmost column in parallel and likewise, the application of the row rule table r rewrites all symbols in the uppermost row. We now compare the generative

$$
M_0 = \dfrac{b\ a}{a\ b} \Rightarrow_{(l/u)} \dfrac{b\ a\ a}{a\ b\ b} \Rightarrow_{(l/u)} \begin{array}{l} b\ a\ a \\ a\ b\ b \\ a\ b\ b \end{array} \Rightarrow_{(l/u)} \begin{array}{l} b\ a\ a\ a \\ a\ b\ b\ b \\ a\ b\ b\ b \end{array} \Rightarrow_{(l/u)} \begin{array}{l} b\ a\ a\ a\ a \\ a\ b\ b\ b\ b \\ a\ b\ b\ b\ b \end{array}
$$

Fig. 3. A sample derivation under (l/u) mode

power of $(l/u)P2DCFL$ with $P2DCFL$.

Theorem 1. *The families of $P2DCFL$ and $(l/u)P2DCFL$ are incomparable but not disjoint, when the alphabet contains at least two symbols.*

Proof. It is clear that the families are not disjoint since the non-trivial picture language of all rectangular picture arrays over $\{a, b\}$ belongs to both of them. In fact the corresponding grammar needs to have only two tables

$$
c = \{a \rightarrow aa, a \rightarrow ab, b \rightarrow ba, b \rightarrow bb\}, r = \left\{a \rightarrow \dfrac{a}{a}, a \rightarrow \dfrac{a}{b}, b \rightarrow \dfrac{b}{a}, b \rightarrow \dfrac{b}{b}\right\}
$$

and axiom pictures a, b.

The picture language L_2 in Example 2 belongs to $(l/u)P2DCFL$ but it cannot be generated by any $P2DCFG$. In fact every column (including the leftmost column) in the picture arrays of L_2 involves the two symbols a, b and only these two. So to generate the picture arrays of L_2 starting from an axiom array, we have to specify column rules for both a, b. The leftmost column will require a column rule that will rewrite b into $ba \cdots a$ and a into $ab \cdots b$ but then the table with these rules can be applied to any other column in a $P2DCFG$. This will result in picture arrays not in the language L_2.

On the other hand the picture language L_1 in Example 1 belongs to $P2DCFL$ but it cannot be generated by any $(l/u)P2DCFG$. In fact there is an unique middle column in every picture array of L_1. Also to the left and right of this middle column there are an equal number of identical columns. Since only the leftmost column can be rewritten in an $(l/u)P2DCFG$, it is not possible to maintain this feature of "equal number of identical columns" if leftmost column rewriting is done.   □

Remark 1. The families $P2DCFL$ and $(l/u)P2DCFL$ coincide if we restrict to only a unary alphabet. Since there is a single symbol and the column rules and the row rules can use only one symbol, rewriting any column is equivalent to rewriting the leftmost column of a picture array.

We now exhibit non-closure of the family $(l/u)P2DCFL$ under the Boolean operations of union and intersection.

Theorem 2. *The family $(l/u)P2DCFL$ is not closed under union.*

Proof. Let $L_1 \subseteq \{a, b, d\}^{**}$ be a picture language such that each $p \in L_1$ of size $(m, n), m \geq 2, n \geq 2$ has the following properties: $p(1, 1) = b$; $p(1, j) = a$, for $2 \leq j \leq n$; $p(i, 1) = a$, for $2 \leq i \leq m$; $p(i, j) = d$, otherwise. Let $L_2 \subseteq \{a, b, e\}^{**}$ be a picture language such that each $p \in L_2$ of size $(r, s), r \geq 2, s \geq 2$ has the following properties: $p(1, 1) = b$; $p(1, j) = a$, for $2 \leq j \leq s$; $p(i, 1) = a$, for $2 \leq i \leq r$; $p(i, j) = e$, otherwise. The languages L_1 and L_2 are generated by $(l/u)P2DCFGs$ G_1 and G_2, respectively. We mention here only the tables of rules and axiom arrays of these grammars. The other components are understood from the tables of rules. The column rule table of G_1 is

$$c_1 = \{b \rightarrow ba, a \rightarrow ad\}$$

while the row rule table is

$$r_1 = \left\{ b \rightarrow \begin{matrix} b \\ a \end{matrix}, a \rightarrow \begin{matrix} a \\ d \end{matrix} \right\}.$$

The column rule table of G_2 is

$$c_2 = \{b \rightarrow ba, a \rightarrow ae\}$$

while the row rule table is

$$r_2 = \left\{ b \rightarrow \begin{matrix} b \\ a \end{matrix}, a \rightarrow \begin{matrix} a \\ e \end{matrix} \right\}.$$

The axiom pictures of G_1 and G_2 are $\begin{smallmatrix} b & a \\ a & d \end{smallmatrix}$ and $\begin{smallmatrix} b & a \\ a & e \end{smallmatrix}$, respectively. Now the union picture language $L_1 \cup L_2$ cannot be generated by any $(l/u)P2DCFG$. In fact, the smallest pictures in $L_1 \cup L_2$ are $\begin{smallmatrix} b & a \\ a & d \end{smallmatrix}$ and $\begin{smallmatrix} b & a \\ a & e \end{smallmatrix}$. Both of these will be the axiom arrays in any $(l/u)P2DCFG$ that could be formed to generate $L_1 \cup L_2$. Also in order to generate the pictures of L_1, column rules of the form $a \to ad \cdots d$ will be required while to generate the pictures of L_2 column rules of the form $a \to ae \cdots e$ will be needed. Likewise for row rules. But then there is no restriction on the application of the tables of rules which will therefore generate pictures not in $L_1 \cup L_2$. □

Theorem 3. *The family $(l/u)P2DCFL$ is not closed under intersection.*

Proof. Let L_s be a picture language consisting of square sized arrays p of the language L_2 in Example 2 i.e. pictures p of size $(n,n), n \geq 2$ with $p(1,1) = b$; $p(1,j) = a$, for $2 \leq j \leq n$; $p(i,1) = a$, for $2 \leq i \leq n$; $p(i,j) = b$, otherwise. We denote here by L_r the picture language L_2 in Example 2 noting that the picture arrays of L_2 are rectangular arrays.

We consider a language L containing of the following three sets of picture arrays:
i) Square arrays with the uppermost row in each array being of the form $xd \cdots d$, the leftmost column of the form $^txe \cdots e$ and with b in all other positions
ii) Rectangular arrays with the uppermost row in each array being of the form $yd \cdots d$, the leftmost column of the form $^tye \cdots e$ and with b in all other positions
iii) the picture arrays of L_s.
    The picture language L_r is generated by the $(l/u)P2DCFG$ of Example 2 while L is generated by an $(l/u)P2DCFG$ G, for which we mention here only the column rule and row rule tables and the axiom array. The column rule tables are

$$c_1 = \{x \to yd, \ e \to eb\}, \ c_2 = \{x \to b, \ e \to a\}.$$

The row rule tables are

$$r_1 = \left\{y \to \begin{smallmatrix} x \\ e \end{smallmatrix}, \ d \to \begin{smallmatrix} d \\ b \end{smallmatrix}\right\}, \ r_2 = \{b \to b, \ d \to a\}.$$

The axiom array is $\begin{smallmatrix} x & d \\ e & b \end{smallmatrix}$. We note that an application of the column rule table c_1 will increase the number of columns by one, after which only the row rule table r_1 can be applied which will then increase the number of rows by one, thereby yielding a square sized array. The application of the tables of rules c_2, r_2 produce the picture arrays in L_s. It is clear that $L_s \subset L$ and $L_s = L_r \cap L$. It can be seen that L_s cannot be generated by any $(l/u)P2DCFG$ (using the alphabet $\{a,b\}$), since the application of the column rule and row rule tables are independent and hence cannot ensure square size of the pictures generated. □

Analogous to $(R)P2DCFG$, we can define a controlled $(l/u)P2DCFG$.

Definition 3. *Let* $G = (\Sigma, P_1, P_2, \{M_0\})$ *be an* $(l/u)P2DCFG$ *G. Let* $\Gamma = P_1 \cup P_2$ *i.e.* Γ *is the set of labels of the column rule and row rule tables of G. Let* $\mathcal{C} \subseteq \Gamma^*$, *whose elements are called control words. The application of the tables in an* l/u *derivation in G is regulated by the control words of* $\mathcal{C}$, *called the control language. An* $(l/u)P2DCFG$ *with a regular and context-free control language is denoted by* $(R)(l/u)P2DCFG$ *and* $(CF)(l/u)P2DCFG$, *respectively. In addition, the family of picture languages generated by* $(R)(l/u)P2DCFGs$ *and* $(CF)(l/u)P2DCFGs$ *is denoted by* $(R)(l/u)P2DCFL$ *and* $(CF)(l/u)P2DCFL$, *respectively.*

It is known [15] that the family of $P2DCFL$ is properly contained in $(R)P2DCFL$. An analogical inclusion holds for the families $(l/u)P2DCFL$ and $(R)(l/u)P2DCFL$.

Theorem 4. $(l/u)P2DCFL \subset (R)(l/u)P2DCFL \subset (CF)(l/u)P2DCFL$.

Proof. The inclusions are straightforward since an $(l/u)P2DCFG$ is an $(R)(l/u)P2DCFG$ on taking the regular control language as Γ^* where Γ is the set of labels of the tables of the $(l/u)P2DCFG$. Also it is well-known [13] that the regular language family is included in the CF family.

The proper inclusion in $(l/u)P2DCFL \subset (R)(l/u)P2DCFL$ can be seen by considering a picture language L_3 consisting of square sized arrays p of the language L_s given in the proof of Theorem 3. This picture language can be generated by the $(l/u)P2DCFG$ G_2 in Example 2 with a regular control language $(cr)^*$. But it is clear that L_3 cannot be generated by an $(l/u)P2DCFG$, since the applications of the column rule and row rule tables are independent.

The proper inclusion of $(R)(l/u)P2DCFL$ in $(CF)(l/u)P2DCFL$ can be shown by considering a picture language L_4 consisting of picture arrays p as in Example 1 but of sizes $(k + 1, 2k + 1), k \geq 1$. The $(CF)(l/u)P2DCFG$ $G^c = (G_4, \Gamma, \mathcal{C})$ generates L_4, where $G_4 = (\Sigma, P_1, P_2, \{M_0\})$ where $\Sigma = \{a, b, e\}$, $P_1 = \{c_1, c_2, c_3\}, P_2 = \{r\}$ with

$$c_1 = \{e \to ea, a \to ab\}, c_2 = \{e \to ae, a \to ba\}, c_3 = \{a \to aa, b \to bb\},$$

$$r = \left\{ e \to \begin{matrix} e \\ a \end{matrix}, a \to \begin{matrix} a \\ b \end{matrix} \right\},$$

$M_0 = \begin{matrix} e\ a \\ a\ b \end{matrix}$ and the tables of rules c_1, c_2, c_3, r are themselves taken as the labels of the corresponding tables, constituting the set Γ. The CF control language is $\mathcal{C} = \{(c_1 r)^n c_2 c_3^n \mid n \geq 0\}$. In order to generate the picture arrays of L_4, the l/u derivations are done according to the control words of $\mathcal{C}$. Starting from the axiom array $M_0 = \begin{matrix} e\ a \\ a\ b \end{matrix}$ the leftmost column of M_0 is rewritten using the column rule table c_1 immediately followed by the row rule table r. This is repeated n times (for some $n \geq 0$) and then the column rule table c_2 is applied once, followed by the application of the column rule table c_3, the same number of times as c_1

followed by r was done, thus yielding a picture array in L_4. But L_4 cannot be generated by any $(l/u)P2DCFG$ with regular control. In fact in a generation of a picture array p in L_4 that makes use of a regular control, if the derivation is generating the part of p to the left of the middle column (made of one e as the first symbol and all other symbols in the column being a's), there will be no information available on the number of columns generated once the derivation "crosses" the middle column, so that the columns to the right of this middle column cannot be generated in equal number. $\qquad\square$

In a $(R)P2DCFG$, the alphabet may contain some symbols called control symbols [2] which might not be ultimately involved in the picture arrays of the language generated. For example, the $(R)(l/u)P2DCFG$ with the $P2DCFG$

$$(\{e,a,b\}, \{c_1,c_2\}, \{r\}, \{\begin{smallmatrix} e & a \\ a & b \end{smallmatrix}\}) \text{ where}$$

$$c_1 = \{e \rightarrow ea, a \rightarrow ab\}, c_2 = \{e \rightarrow a, a \rightarrow a\},$$

$$r = \left\{ e \rightarrow \begin{smallmatrix} e \\ a \end{smallmatrix}, a \rightarrow \begin{smallmatrix} a \\ b \end{smallmatrix} \right\},$$

and the control language $\{(c_1 r)^n c_2 | n \geq 0\}$ generates picture arrays p such that the uppermost row and the leftmost column of p involve only the symbol a while all other positions have the symbol b. But the alphabet contains a symbol e which ultimately does not appear in the picture arrays of the language. Such a symbol is referred to as a control symbol or a control character in the context of an $(R)(l/u)P2DCFG$. A picture language L_d is considered in [2] given by $L_d = \{p \in \{a,b\}^{++} |\ |p|_{col} = |p|_{row}, p(i,j) = b, \text{ for } i = j, p(i,j) = a \text{ for } i \neq j\}$ and is shown to require at least two control symbols to generate it using a $P2DCFG$ and a regular control language.

Lemma 1. *[2] The language L_d cannot be defined by using less than two control characters and a P2DCFG with a regular control language.*

We show in the following Lemma that in an $(R)(l/u)P2DCFG$ the picture language L_d can be generated with a single control character.

Lemma 2. *The language L_d can be defined by an $(R)(l/u)P2DCFG$ that uses a single control character. Moreover, L_d is not in $(l/u)P2DCFL$.*

Proof. The $(R)(l/u)P2DCFG$ with the $(l/u)P2DCFG$ given by

$$(\{0,1,2\}, \{c\}, \{r\}, \{\begin{smallmatrix} 1 & 0 \\ 0 & 1 \end{smallmatrix}\}) \text{ where}$$

$$c = \{1 \rightarrow 12, 0 \rightarrow 00\}, r = \left\{ 1 \rightarrow \begin{smallmatrix} 1 \\ 0 \end{smallmatrix}, 2 \rightarrow \begin{smallmatrix} 0 \\ 1 \end{smallmatrix}, 0 \rightarrow \begin{smallmatrix} 0 \\ 0 \end{smallmatrix} \right\},$$

and control language $(cr)^*$ generates L_d. Here, 2 is the only control character. It is clear that if there are only two symbols $0, 1$ in the alphabet, then, for example, there need to be two column rules $0 \rightarrow 01, 0 \rightarrow 00$ in a table to maintain the diagonal of 1's but this will yield pictures not in L_d. A similar reason holds for row rules. This shows L_d cannot be in $(l/u)P2DCFL$. $\qquad\square$

Finally, we compare $(l/u)P2DCFL$ with the class LOC [4] of local picture languages whose pictures are defined by means of tiles i.e. square pictures of size $(2,2)$.

Theorem 5. *The families $(l/u)P2DCFL$ and LOC are incomparable but their intersection is not empty.*

Proof. The picture language of all rectangular arrays over a one letter alphabet $\{a\}$ is clearly in $(l/u)P2DCFL$ and is also known [2] to be in LOC. But the language of rectangular pictures with an even number of rows and an even number of columns is not in LOC [2] but is in $P2DCFL$ [2] and hence in $(l/u)P2DCFL$, by Remark 1. On the other hand, the language L_d in Lemma 2 is in LOC [2] but again by Lemma 2, L_d is not in $(l/u)P2DCFL$.   □

4   Conclusion

A variant of $P2DCFG$ [14,15] rewriting only the leftmost column or the uppermost row of a picture array is considered and properties of the resulting family $(l/u)P2DCFL$ of picture languages are obtained. Properties such as closure or non-closure under row or column concatenation of arrays or membership problem and others remain to be investigated. It will also be of interest to allow erasing rules of the form $a \to \lambda$ and examine the effect of using these rules in the derivations of the picture arrays.

Acknowledgements.   This   work   was   supported   by   the   Departament d'Economia i Coneixement, Generalitat de Catalunya and by the European Regional Development Fund in the IT4Innovations Centre of Excellence project (CZ.1.05/1.1.00/02.0070) and TAČR TE01010415 grant. The authors thank the anonymous referee for useful comments regarding the first version of this paper.

References

1. Bersani, M.M., Frigeri, A., Cherubini, A.: On some classes of 2D languages and their relations. In: Aggarwal, J.K., Barneva, R.P., Brimkov, V.E., Koroutchev, K.N., Korutcheva, E.R. (eds.) IWCIA 2011. LNCS, vol. 6636, pp. 222–234. Springer, Heidelberg (2011)
2. Bersani, M.M., Frigeri, A., Cherubini, A.: Expressiveness and complexity of regular pure two-dimensional context-free languages. Int. J. Comput. Math. 90, 1708–1733 (2013)
3. Fernau, H.: Regulated grammars under leftmost derivation. Grammars 3, 37–62 (2000)
4. Giammarresi, D., Restivo, A.: Two-dimensional languages. In: Handbook of Formal Languages, vol. 3, pp. 215–267. Springer (1997)
5. Maurer, H.A., Salomaa, A., Wood, D.: Pure grammars. Inform. and Control 44, 47–72 (1980)

6. Masopust, T., Techet, J.: Leftmost derivations of propagating scattered context grammars: a new proof. Discrete Math. and Theoretical Comp. Sci. 10, 39–46 (2008)
7. Meduna, A.: Automata and Languages: Theory and Applications. Springer, London (2000)
8. Meduna, A., Zemek, P.: One-sided random context grammars with leftmost derivations. In: Bordihn, H., Kutrib, M., Truthe, B. (eds.) Languages Alive. LNCS, vol. 7300, pp. 160–173. Springer, Heidelberg (2012)
9. Nagy, B.: Derivation trees for context-sensitive grammars. In: Automata, Formal Languages and Algebraic Systems (AFLAS 2008), pp. 179–199. World Scientific Publishing (2010)
10. Rosenfeld, A.: Picture Languages. Academic Press, Reading (1979)
11. Rosenfeld, A., Siromoney, R.: Picture languages—a survey. Languages of Design 1, 229–245 (1993)
12. Rozenberg, G., Salomaa, A. (eds.): Handbook of Formal Languages, vol. 1–3. Springer, Berlin (1997)
13. Salomaa, A.: Formal Languages. Academic Press, Reading (1973)
14. Subramanian, K.G., Ali, R.M., Geethalakshmi, M., Nagar, A.K.: Pure 2D picture grammars and languages. Discrete Appl. Math. 157, 3401–3411 (2009)
15. Subramanian, K.G., Nagar, A.K., Geethalakshmi, M.: Pure 2D picture grammars (P2DPG) and P2DPG with regular control. In: Brimkov, V.E., Barneva, R.P., Hauptman, H.A. (eds.) IWCIA 2008. LNCS, vol. 4958, pp. 330–341. Springer, Heidelberg (2008)
16. Subramanian, K.G., Rangarajan, K., Mukund, M. (eds.): Formal Models, Languages and Applications. Series in Machine Perception and Artificial Intelligence, vol. 66. World Scientific Publishing (2006)
17. Wang, P.S.-P. (ed.): Array Grammars, Patterns and Recognizers. Series in Computer Science, vol. 18. World Scientific Publishing (1989)

Discovering Features Contexts
from Images Using Random Indexing

Haïfa Nakouri[1] and Mohamed Limam[1,2]

[1] Institut Supérieur de Gestion, LARODEC Laboratory
University of Tunis, Tunisia
[2] Dhofar University, Oman
nakouri.hayfa@gmail.com, mohamed.limam@isg.rnu.tn

Abstract. Random Indexing is a recent technique for dimensionality reduction while creating Word Space model from a given text. The present work explores the possible application of Random Indexing in discovering feature semantics from image data. The features appearing in the image database are plotted onto a multi-dimensional Feature Space using Random Indexing. The geometric distance between features is used as an indicative of their contextual similarity. K-means clustering is used to aggregate similar features. In this paper, we show that the Feature Space model based on Random Indexing can be used effectively to constellate similar features. The proposed clustering approach has been applied to the Corel databases and motivating results have obtained.

Keywords: Indexing, Geometric Distance, K-means, Clustering.

1 Introduction

Most of the image analysis approaches consider each image as a whole, represented by a D-dimensional vector. However, the user's query is often just one part of the query image (i.e. a region in the image that has an obvious semantic meaning). Therefore, rather than viewing each image as a whole, it is more reasonable to view it as a set of semantic regions of features. In this context, we consider an image feature as a relevant semantic region of an image that can summarize the whole or a part of the context of the image.

In this work, we propose the Feature Space model similarly to the Word Space model [13] that has long been used for semantic indexing of text. The key idea of a Feature Space model is to assign a vector (generally a sparse vector) to each feature in the high dimensional vector space, whose relative directions are assumed to indicate semantic similarities or similar representations of the features. However, high dimensionality of the semantic space of features, sparseness of the data and large sized data sets are the major drawbacks of the Feature Space model.

Random Indexing (RI) [7,12] is an approach developed to cope with the problem of high dimensionality in the Word Space model and many works used it for text indexing and words' semantic creation [2,5,11,14,15,17]. It is an incremental

R.P. Barneva, V.E. Brimkov, and J. Šlapal (Eds.): IWCIA 2014, LNCS 8466, pp. 134–145, 2014.

approach proposed as an alternative to Latent Semantic Indexing (LSI) [8]. Wan et al. [16] used RI to identify and capture Web users' behaviour based on their interest-oriented actions. To the best of our knowledge, no Random Indexing approaches have been used to deal with image features in the Feature Space model especially for similar semantics discovery between features in image data sets. In this paper we aim to show that a Feature Space model constructed using Random Indexing can be used efficiently to cluster features, which in turn can be used to identify the representation or the context of the feature. In a Feature Space model, the geometric distance between the features is an indicative of their semantic similarity.

The interesting point in RI is that it enables finding relevant image documents even if they do not contain the query key features and the whole procedure is incremental and automatic. The fact that RI does not require an exact match to return useful results fits perfectly with the scenario of feature image clustering. Assume that there is a query image of a 'cat' in the grass and the user is interested in finding all images in the database that contain a 'cat'. It is obviously not a good idea to use exact match since no 'cat' image would have exactly the same low-level features with the query image itself. Hence, in the context of our work, if we consider an image as a document, the 'cat' object is then one of the words in the document. The only difference is that the 'cat' object is not a word but a multidimensional feature vector. The objective of using the *context vectors* computed on the language data is to map the features onto the Feature Space.

We used K-mean [9] method to agglomerate similar features and each constellation represents a context of images.

In this paper, we attempt to show that the Feature Space model based on Random Indexing can be used efficiently to cluster features, which in turn can be used to approximate the context represented by a feature.

The rest of this paper is organized as follows. Sect. 2 introduces the Feature Space model and the Random Indexing approach. Sect. 3 presents the proposed feature clustering process based on Random Indexing. Sect. 4 presents the experimental results of the proposed work.

2   Vector-Based Feature Analysis Using Random Indexing

Basically, vector-based semantic analysis is a technology for extracting semantically similar terms from textual data by observing the distribution and collocation of terms in text. The result of running a vector-based semantic analysis on a text collection can be used to find correspondences across terms. The meaning or representation of a term is interpreted by the context it is used in. By analogy to the Word Space model which is a spacial representation of word meaning, we consider a Feature Space model as a spacial representation of feature meaning.

2.1   Feature Space Model

In this model, the complete features of any image (containing n features) can be represented in a n-dimensional space in which each feature occupies a specific point in the space, and has a vector associated with it defining its meaning. The features are placed on the Feature Space model according to their distributional properties in the image, such that:

1. The features that are used within similar group of features (i.e. in a similar context) should be placed nearer to each other.
2. The features that lie closer to each other in the Feature Space represent the same context. Meanwhile, the features that lie farther from each other in the Feature Space model are dissimilar in their representation.

2.2 The Feature Space Model and Random Indexing

Random Indexing (RI) is based on Kanerva's work [6] on sparse distributed memory. It was proposed by Karlgren and Sahlgren [7,12] and was originally used as a text mining technique. It is a word-occurrence based approach to statistical semantics. RI uses statistical approximations of the full word-occurrences data to achieve dimensionality reduction. Besides, it is an incremental vector space model that is computationally less demanding. The Random Indexing model reduces dimensionality by, instead of giving each word a whole dimension, it gives them a random vector with less dimensionality than the total number of words in the text. Thus, RI results in a much quicker time and fewer required dimensions.

Random Indexing used sparse, high-dimensional random *index vectors* to represent image features. Sparsity ensures that the chance of any two arbitrary index vectors having an overlapping meaning (i.e. a cosine similarity [13] that is non-zero) is very low. Given that each feature has been assigned a random *index vector*, features similarities can be calculated by computing a feature-context co-occurrence matrix. Each row in the matrix represents a feature and the feature vectors are of the same dimensionality as are the random vectors assigned to images. Each time a feature is found in an image, that image's random *index vector* is added to the row of the feature in question. In this way, features are represented in the matrix by high-dimensional semantic *context vectors* which contain trances of each context the feature has been observed in. The underlying assumption is that semantically similar features should be possible to compute the semantic similarity between any given features by calculating the similarity between their *context vectors*. Mathematically, this is done by computing the cosine of the angles between the *context vectors* [13]. Generally, to measure the divergence between vectors, the angle between the vectors is used, and cosine of the angle can then be used as the numeric similarity (cosine is equal to 1 for identical vectors and 0 for orthogonal vectors). Thus, this cosine similarity measure will reflect the distributional (or contextual) similarity between features (see Sect. 3).

This technique is akin to Latent Semantic Analysis (LSA) of Indexing (LSI) [8], except that no dimension reduction (e.g. Singular Value Decomposition (SVD)) is needed to reduce the dimensions of the co-occurrence matrix, since the dimensionality of the random *index vectors* is smaller than the number of images in the training data. This makes the technique more efficient that the LSI methods, since SVD is a computationally demanding operation. The technique is also easier to scale and more flexible as regards unexpected data than are methods which rely on dimension reduction. A new image does not require a larger matrix but will simply be assigned a new random *index vector* of the same dimensionality as the preceding ones and a new term requires no more than a row in the matrix.

The size of the context used to accumulate feature-image matrix may range from just few features on each side of the focus feature to the entire image data consisting of more than hundred features [12].

In the context of our work, the context of a feature image is understood as the visual surrounding of a feature. For instance, an "umbrella" and "surf board" are two features representing the context "beach".

A context vector thus obtained can be used to represent the distributional information of the feature into the geometrical space. This is similar to each feature being assigned a unique unary vector of dimension d, called *index vector*. *The context vector* for a feature can be obtained by summing up the *index vectors* of the features on the either side of it. In other words, all the features representing the same context should have approximately equal *context vectors*.

Random Indexing accumulates *context vectors* to form the Feature Space model in a two step process.

1. Each feature in the image is assigned a unique and randomly generated vector called the *index vector*. Each feature is also assigned an initially empty *context vector* which has the same dimensionality (d) as the *index vector*. A unique d-dimensional *index vector* is assigned and randomly generated to each feature. These *index vectors* are sparse, high dimensional, and ternary (i.e. 1, -1, 0). In other words, the dimensionality (d) is in the order of hundreds, and that they consist of all small numbers (ϵ) of randomly distributed $+1$ and -1, with the remaining elements of the vector to 0. In our work, we allocate each elements as follows:

$$\begin{cases} +1 \text{ for the probability } (\epsilon/2)/d \\ 0 \text{ for the probability } (d - \epsilon/d \\ -1 \text{ for the probability } (\epsilon/2)/d \end{cases}$$

2. The *context vectors* are then accumulated by advancing through the image data set one feature taken at time, and then adding the context's *index vector* to the focus feature's *context vector*. When the entire data is processed, the d-dimensional *context vectors* are effectively the sum of the feature's contexts. *Context vectors* are produced by scanning through the images. As scanning the image data, each time a feature occurs in a context, that context's d-dimensional *index vector* is added to the *context vector* of the feature. This way, features are represented by d-dimensional *context vectors* that are the sum of the *index vectors* of all the contexts in which the feature appears.

In further experiments, we used Random Indexing to index aligned features and extract semantically similar features across image documents. Compared to other vector space methodologies, we find Random Indexing unique in the following way. First, it is an incremental method, which means that the *context vectors* can be used for similarity computations even after just few examples have been encountered. By contrast, most other vector space models (e.g. LSA) require the entire data to be sampled before similarity computations can be performed. Second, the dimensionality d of the vectors is a parameter in Random Indexing. This means that d does not change once it has been set. In fact, new data increases the values of the elements of the *context vectors*, but never their dimensionality. Increasing dimensionality can lead to significant scalability problems in other vector space methods. Third, Random Indexing uses implicit dimension reduction, since the fixed dimensionality d is much lower than the number of contexts in the data. This leads to a significant gain in processing time and memory consumption as compared to vector space methods that employ computationally expensive algorithms (e.g. SVD). Finally, Random Indexing can be used with any type of context and by this work we extend it to another type of data which is image data instead of text data.

3 Feature Clustering

Figure 1 illustrates the overall procedure of the feature clustering process based on Random Indexing. The clustering procedure is based on three steps: data preprocessing, modelling the Feature Space using Random Indexing and the feature clustering. More details are outlined in this Section.

3.1 Data Preprocessing

The preprocessing phase consists in the feature extraction from images. To this end, we first need to perform an image segmentation and then extract the relevant features. In our experiment, we choose to use the conventional Blob-world [1] as our image segmentation method. Figure 2 shows an example of segmented images using the Blob-world method and the extracted features.

3.2 Feature Space Model Using Random Indexing

Once all relevant features are extracted, further analysis should be done to find common contexts between features and create a proper context model for the features clustering. Feature semantics are computed by scanning the features set and keeping a running sum of all *index vectors* for the features that co-occur. Actually, a link exists between the occurrence of a feature and its semantics. Finally, the set of generated context vectors represent the Feature Space model corresponding to the image data set. Algorithm 1 summarizes the *context vectors* generation procedure.

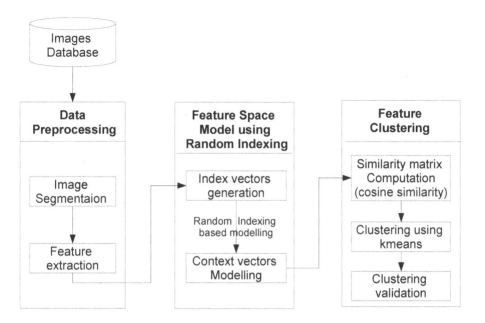

Fig. 1. Feature clustering approach based on Random Indexing

Fig. 2. Examples of segmented images

3.3 Similarity Measure in the Feature Space Model

Basically, *context vectors* give the location of the word in the Word Space. Similarly, we can assume that *context vectors* give the location of the feature in the Feature Space model. In order to determine how similar the features are in the

---

Algorithm 1: Context vector generation

INPUT: Features f_i, $i = 1, \ldots, n$.
OUTPUT: $n \times d$ context window A.

1. For each feature f_i, obtain a d-dimensional *index vector* $ind_i, i = 1, \ldots, n$ where n is the total number of features.
2. Scanning the feature set, for each feature f_i appearing in the same context than another feature, update its context's vector c_i by adding the feature's corresponding ind_i.
3. Create the feature-to image ($n \times d$) matrix, also called the context window, where each row is the context vector c_i of each single feature.

---

context, a similarity measure has to be defined. Various schemes e.g. scalar product or vector, Euclidean distance, Minkowski metrics [12], are used to compute similarity between vectors corresponding to the features. However, the cosine distance [12] might make sense for these data because it would ignore absolute sizes of the measurements, and only consider their relative sizes. Thus, two flowers that were different sizes, but which had similarly shaped petals and sepals, might not be close with respect to squared Euclidean distance, but would be close with respect to cosine distance. We have used cosine of the angles between pairs of vectors x_i and y_i , $i = 1, \ldots, n$ of the context window A generated with Random Indexing to compute normalized vector similarity. The cosine angle between two vectors x and y is defined as:

$$sim_\propto(x, y) = \frac{xy}{abs(x)abs(x)} = \frac{\sum_{i=1}^{n} x_i y_i}{\sqrt{(\sum_{i=1}^{n} x_i^2)}\sqrt{(\sum_{i=1}^{n} y_i^2)}} \tag{1}$$

The cosine measure is a frequently used similarity metric in vector space research. The advantage of the cosine metric over other metrics is that it provides a fixed measure of similarity, which ranges from 1 (for identical vectors) to 0 (for orthogonal vectors) and -1 (for vectors pointing in the opposite directions). Moreover, it is comparatively efficient to compute. Hence, a similarity index between different features is defined, and a similarity matrix is generated from the features appearing in the images data set. Each cell of the similarity matrix contains the numerical value of the similarity between any pair of features.

3.4   Feature Clustering

The third phase of the clustering process takes in as input the similarity matrix between features. These objects have a cosine similarity between them, which can be converted to a distance measure, and then be used in any distance based classifier, such as nearest neighbor classification. A simple application of the K-means algorithm is performed to cluster the features. K-means [9] is a popular and conventional clustering algorithm that aims to partition n observations into k clusters in which each observation belongs to the cluster with the nearest mean.

Fig. 3. Example images from the Corel database

This partition-based clustering approach has been widely applied for decades and should be suitable for our clustering problem.

4   Experiments

4.1   The Training Dataset

In this work, we used the Corel database [10] for training and constructing our Feature Space model. The Corel image database contains close to 60,000 general purpose photographs. A large portion of images in the database are scene photographs. The rest includes man-made objects with smooth background, fractals, texture patches, synthetic graphs, drawings, etc. This database was categorized into 599 semantic concepts. Each concept/category/context, containing roughly 100 images, e.g. 'landscape', 'mountain', 'ice', 'lake', 'space', 'planet', 'star'. For clarification, general-purpose photographs refer to pictures taken in daily life in contrast to special domain such as medical or satellite images. Unfortunately, for copyright restrictions, we only have access to 1,000 images from the coral DB for which we can distinguish 10 different contexts. Thus we can expect 10 well separated clusters from the proposed RI-based clustering. Figure 3 shows some image examples from the Corel database.

4.2   Clustering Validity Measures

In order to evaluate the performance of the proposed clustering algorithm, we use the CS index [4] that computes the ratio of *Compactness* and *Separation*.

A common measure of *Compactness* is the intra-cluster variance within a cluster, named *Comp*:

$$Comp = \frac{1}{k} \sum_{i=1}^{k} \| \gamma(C_i) \|, \tag{2}$$

where $\gamma(X)$ represents the variance of data set X. *Separation* is computed by the average of distances between the centers of different clusters:

$$Sep = \frac{1}{k} \sum \| z_i - z_j \|^2, \qquad i = 1, 2, \ldots, k-1, \qquad j = i+1, \ldots, k \tag{3}$$

It is clear that if the data set contains compact and well separated clusters, the distance between the clusters is expected to be large and the diameter of the clusters is expected to be small. Thus, cluster results can be compared by taking the ratio between *Comp* and *Sep*:

$$CS = \frac{Comp}{Sep}. \tag{4}$$

Based on the definition of CS, we can conclude that a small value of CS indicates compact and well-separated clusters. CS reaches its best score at 0 and worst value at 1. Therefore, the smaller it is the the better the clusters are formed.

4.3   Parameter Settings

We experimented the impact of some key parameters and assigned initial values to them.

Dimensionality: in order to evaluate the effects of varying dimensionality on the performance of Random Indexing in our work, we computed the values of CS with d ranging from 100 to 600. The performance measures are reported using average values over 5 different turns. Figure 4 depicts the results and shows that for $d = 300$ we get the smallest CS value. Therefore, we choose $d = 300$ as the dimension of the *index vectors* for Random Indexing, which is way less than the original $D = 1000$ (corresponding to total number of images in the data set). As stated in [12], even though the low-dimensional vector space is still relatively large (a few hundred dimensions), it is nevertheless lower than the original space corresponding to the data size (thousands of dimensions).

Sparsity of the index vectors: another parameter is crucial for the quality of our indexing is the number of +1, and −1 in the index vectors ϵ. We use $\epsilon = 10$ as proposed in [3].

4.4   Clustering Results

As indicated above, the data consists of 1000 features and 10 contexts. For the clustering results, the 10 predicted clusters corresponding to the 10 different

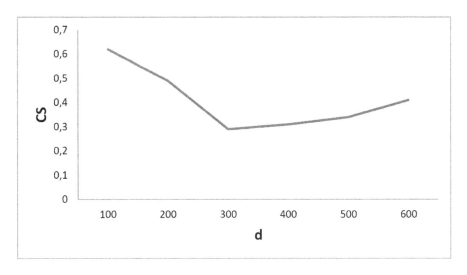

Fig. 4. The impact of different d values to RI-based feature clustering

contexts have been correctly formed and Table 1 shows some of the formed clusters/contexts and their assigned features.

Table 1. Some of the features and their discovered contexts

Description of the feature cluster/Context	
Beach umbrella	Beach
Person lying in the beach	Beach
The Colosseum	Landmarks
London bus	Vehicle

We report the rest of the results using three other validation criteria: precision, recall and the F-measure. These three measures are widely used in pattern recognition and information retrieval. According to our evaluation context, we slightly changed the definitions: *Precision* of a feature is defined as the ratio of the correct clusters to the total number of clusters it is assigned to. The precision (P) of a the algorithm is the average precision of all features. *Recall* of a feature is defined as the ratio of the correct clusters of the feature and the total number of contexts the feature is used in the data set. The recall (R) of the algorithm is the average recall of the features. P and R range between 0 and 1. The F-measure (F) is the combination result of precision and recall and is given by:

$$F = \frac{2RP}{R + P}. \tag{5}$$

The F-measure reaches its best value at 1 and worst score at 0. As showed in Table 2, the best results of the proposed measures are given for dimension

$d = 300$: the smallest Compactness Separation ($CS = 0.294$) and accordingly the largest F-measure ($F = 0.701$). The best formed clusters (e.g. with the least CS index) cause a decrease in precision and hence in F-measure. It can be observed from the results that Random Indexing can improve the quality of features clustering and allows the construction of a high quality Feature Space model. For all context discoveries, a feature is assigned to a cluster if its closer to this cluster's center. Thus, a feature is assigned to its most similar context.

Nevertheless, in real applications, a single feature can represent more than one context, in other words, a feature is not assigned to exactly one cluster. For instance, the 'London bus' can represent both the 'vehicle' and 'London' contexts and 'The Colosseum' feature can be clustered in both the 'Landmarks' or 'Italy' contexts. Thus, we can consider in a future work the idea of fuzzy clustering where each feature can be assigned to more than one cluster, probably with varying degrees of membership.

Table 2. Results of RI-based clustering

Dimension	d=200				d=300				d=400			
Validation Measure	CS	P	R	F	CS	P	R	F	CS	P	R	F
RI-Clustering	0.49	0.590	0.324	0.418	0.29	0.821	0.612	0.701	0.31	0.683	0.542	0.604

5   Conclusion

In this paper, we have used a Random Indexing based approach, in conjunction with the K-means clustering technique to discover feature semantics from images. The approach works efficiently on the Corel database. However, the proposed approach does not consider the possibility of assigning a feature to more than just one cluster. We will focus on this issue in further works in order to measure the efficiency of the proposed clustering schema on a wide perspective.

References

1. Carson, C., Belongie, S., Greenspan, H., Malik, J.: Blobworld: Image Segmentation Using Expectation-Maximization and Its Application to Image Querying. IEEE Trans. on Pattern Analysis and Machine Intelligence 24(8), 1026–1038 (2002)
2. Giesbrecht, E.: In Search of Semantic Compositionality in Vector Spaces. In: Rudolph, S., Dau, F., Kuznetsov, S.O. (eds.) ICCS 2009. LNCS, vol. 5662, pp. 173–184. Springer, Heidelberg (2009)
3. Gorman, J., Curran, J.R.: Random Indexing using Statistical Weight Functions. In: Proceedings of EMNLP, pp. 457–464 (2006)
4. Halkidi, M., Vazirgiannis, M., Batistakis, Y.: Quality Scheme Assessment in the Clustering Process. In: Zighed, D.A., Komorowski, J., Żytkow, J.M. (eds.) PKDD 2000. LNCS (LNAI), vol. 1910, pp. 265–276. Springer, Heidelberg (2000)
5. Hare, M., Jones, M., Thomson, C., Kelly, S., McRae, K.: Activating event knowledge. Cognition Journal 111(2), 151–167 (2009)

6. Kanerva, P.: Sparse Distributed Memory and Related Models. In: Associative Neural Memories, pp. 50–76. Oxford University Press (1993)
7. Karlgren, J., Sahlgren, M.: From words to understanding. In: Uesaka, Y., Kanerva, P., Asoh, H. (eds.) Foundations of Real-world Intelligence, pp. 294–308 (2001)
8. Landauer, T.K., Foltz, P.W., Laham, D.: An Introduction to Latent Semantic Analysis. In: 45th Annual Computer Personnel Research Conference – ACM (2004)
9. MacQueen, J.: Some Methods for Classification and Analysis of Multivariate Observations. In: Proceedings of the 5th Berkeley Symposium on Mathematical Statistics and Probability, pp. 281–297 (1967)
10. Müller, H., Marchand-Maillet, S., Pun, T.: The Truth about Corel - Evaluation in Image Retrieval. In: Lew, M., Sebe, N., Eakins, J.P. (eds.) CIVR 2002. LNCS, vol. 2383, pp. 38–49. Springer, Heidelberg (2002)
11. Chatterjee, N., Mohan, S.: Discovering Word Senses from Text Using Random Indexing. In: Gelbukh, A. (ed.) CICLing 2008. LNCS, vol. 4919, pp. 299–310. Springer, Heidelberg (2008)
12. Sahlgren, M.: An Introduction to Random Indexing. In: Methods and Applications of Semantic Indexing Workshop at the 7th International Conference on Terminology and Knowledge Engineering, TKE (2005)
13. Sahlgren, M.: The Word-Space Model: Using Distributional Analysis to Represent Syntagmatic and Paradigmatic Relations Between Words in High-Dimensional Vector Spaces. Ph.D. dissertation, Department of Linguistics, Stockholm University (2006)
14. Turian, J., Ratinov, L., Bengio, Y.: Word Representations: A Simple and General Method for Semi-supervised Learning. In: Proceedings of the 48th Annual Meeting of the Association for Computational Linguistics, pp. 384–394 (2010)
15. Turney, P.D., Pantel, P.: From Frequency to Meaning: Vector Space Models of Semantics. J. Artif. Int. Res. 37(1), 141–188 (2010)
16. Wan, M., Jönsson, A., Wang, C., Li, L., Yang, Y.: Web user clustering and Web prefetching using Random Indexing with weight functions. Knowledge and Information Systems 33(1), 89–115 (2012)
17. Widdows, D., Ferraro, K.: Semantic vectors: a scalable open source package and online technology management application. In: Proceedings of the Sixth International Language Resources and Evaluation (LREC 2008), pp. 1183–1190 (2008)

Decomposition of a Bunch
of Objects in Digital Images

Pavel Štarha and Hana Druckmüllerová

Institute of Mathematics, Faculty of Mechanical Engineering,
Brno University of Technology
Technická 2, 616 69 Brno, Czech Republic
starha@fme.vutbr.cz, ydruck00@stud.fme.vutbr.cz

Abstract. In object recognition methods, we often come across a task
to identify objects that originated from several overlapping elementary
objects. After image segmentation, this bunch of elementary objects is
identified as one object, which has completely different geometrical properties from the elementary objects. The information about the number
of objects, their size and shape is degraded. The paper considers a step
towards solving this problem. It describes a mathematical method for
decomposition of objects formed by slightly overlapping / touching objects in separate elementary objects. The method is based only on the
geometrical properties of the object of interest and on the assumption of
the shape of the elementary objects.

Keywords: Segmentation, Geometric properties, Overlapping Objects,
Recrystallization.

1   Introduction

A frequent task in image processing is analyzing objects, their geometrical and
colorimetric properties, and their distribution in the image (number, density). A
crucial step in the processing is objects' identification by segmenting the image
into the objects and the background. Various numerical methods are used for
image segmentation, they can also take into consideration the properties of the
objects to be detected. The most common methods of image segmentation are
thresholding and detection of object boundary. None of the segmentation methods can be ideal and universal. If the objects lie close to each other or they even
slightly overlap, most segmentation methods do not separate these objects and
instead of several elementary objects, one large object is identified, which has
completely different geometrical properties from the original elementary objects.
This also strongly affects the number of detected objects. In some applications,
objects may overlap completely or partially — then the identification of objects
is very complicated if no additional information is provided (e.g., the shape of
the objects). In this contribution, we deal with the case when the objects touch
each other or overlap only slightly. We propose a method that separates the
objects in more parts.

R.P. Barneva, V.E. Brimkov, and J. Šlapal (Eds.): IWCIA 2014, LNCS 8466, pp. 146–157, 2014.

There have been several studies published on this topic. Their major difference between most of these studies and our contribution is that they use more information than just the segmented image. Their input images are gray-scale images or they assumed a special shape of the elementary objects. For example, Cheng and Rajapakse[2] search for the centers of the objects defined as the points at highest distance from the object boundary (by means of the adaptive H-minima transform); then they use a suitable metric to determine the boundary between the objects. Bai, Sun, and Zhou[1] detect the boundary, then its concave points, and fit ellipses in the object. This method is based on the assumption that objects have elliptic shape, therefore a bunch of overlapping objects have the shape of the union of the ellipses. The method is in some sense similar to the method of Kumar et al. [4], who take the initial points for the construction of new boundaries between the overlapping objects as concavities in the object boundary. The next step is optimization of the new boundary based on several criteria. Svensson [9] proposed a method that is suitable for faint objects with fuzzy boundaries such as partially transparent cells. The method is based on the reverse fuzzy distance transform. Finally, Vincent and Soille [10] simulate flooding of the image using a queue of pixels. This method works on gray-scale images.

Our method proposed in this contribution requires a segmented image, in fact a one-bit image. We do not suppose any special shape of the elementary objects, only the geometry of the existing objects is used. The idea of the method is based on the growth of crystals. All crystals have the same growth rate and differ in the time when their growth starts.

In Section 2 we describe the standard method of image segmentation by thresholding, which is the core of the method proposed later in our contribution. In Section 3, we propose the method of recrystallization for splitting a bunch of elementary objects into elementary objects. It is first theoretically derived in the continuous case in Section 3.1 and then the practical implementation for images with discrete pixels is shown in Section 3.2. Finally, in Section 4 we conclude with some remarks.

2   Segmentation by Thresholding

In this section, we focus on the standard method of image segmentation by thresholding, which is the core of many more sophisticated image segmentation methods [7], [6] as well as the method of recrystallization proposed in this contribution. The thresholding method marks each pixel as an object pixel or a background pixel based on its colorimetric properties (brightness, hue, saturation). For a method based on hue thresholding see [8]. The decision criteria in the standard thresholding segmentation does not depend on the neighborhood of the pixel to be decided about. Figure 1 shows an example of image segmentation by means of thresholding. It shows separate elementary objects as well as a bunch of overlapping objects. The standard method of image segmentation by thresholding analyzes the bunch as one object, which has completely different

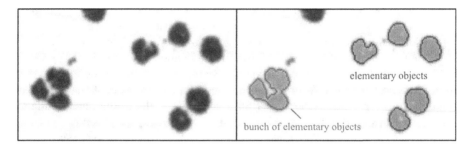

Fig. 1. Image segmentation by thresholding – example of touching elementary objects

geometrical properties from the elementary objects. In this situation, we obtain incorrect information about the studied objects. In the case that the bunch is composed of objects that only touch each other, the information about the overall area of the objects is preserved, to the contrary of partially overlapping objects as it is shown in Figure 2. It is impossible to calculate the area occupied by the objects in this case. To find out the number of elementary objects, we

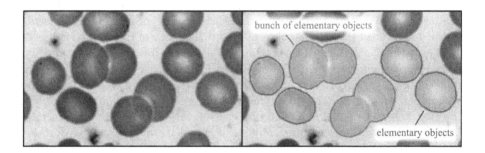

Fig. 2. Partially overlapping elementary objects

have to split the bunches of overlapping objects in a suitable way and also to reconstruct the shape and size of the objects in the overlapping parts if we need to know the area occupied by the objects as well.

3   The Recrystallization Method

In this section we propose a method of separation of a bunch of overlapping objects in more parts. The methods is based on recrystallization, which means that objects are reconstructed from their 'crystallization nuclei'. In our case, it is a point which is the furthest from the object boundary or its part. The object gradually grows from its nucleus. There might be more nuclei in one object. As soon as two growing cells of the object belonging to two different nuclei interact, a boundary of these cells is formed.

The method and its background will be first described in the continuous case, i.e. in the Euclidean plane $\mathbb{E}^2$ and then the solution for the discrete case will be presented.

3.1   The Continuous Description

Note 1. The boundary of a set $A \subset \mathbb{E}^2$ will be denoted by ∂A and its interior by $\text{int}\, A$. $\subset$ means a subset, no matter if the subset is proper or not. $k(X, r)$ will denote an open circle with center in X and radius r. Its boundary, i.e. a circle as a curve will be $\partial k(X, r)$, a closed circle will be $\overline{k}(X, r)$. $\rho(X, Y)$ will stand for the standard Euclidean metric on $\mathbb{E}^2$. In the text below, M will always be a compact subset of $\mathbb{E}^2$ with a non-empty connected interior.

For more about these concepts see [5].

Definition 1. *Let there be an $r > 0$ such that there is at least one point X such that an open ball with center in X and radius r fulfills the condition*

$$k(X, r) \subset \text{int}\, M \;\wedge\; \overline{k}(X, r) \cap \partial M \neq \emptyset. \tag{1}$$

The set equal to the union of all closed balls $\overline{k}(X, r)$ complying with condition (1) is called the r-embedding of set M and is denoted $\widehat{M_r}$.

Figure 3 shows examples of r-embedding $\widehat{M_r}$ for two values of $r > 0$. A curve

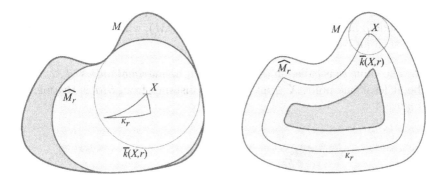

Fig. 3. Examples of r-embedding with radius r and curve κ_r

κ_r of points that have a constant distance r from the boundary ∂M of set M is called a *curve parallel with ∂M*. The center X of a circle $k(X, r)$ from the previous definition belongs to its curve parallel with ∂M.

Theorem 1. *Let $X \in \text{int}\, M$ arbitrarily. Then for every circle $k(X, r)$ such that condition (1) is fulfilled for a suitable r it holds*

$$r = \varrho(X, \partial M).$$

Proof. Condition (1) can be rewritten as

$$\text{int}\, k(X,r) \subset \text{int}\, M \,\wedge\, \partial k(X,r) \cap \partial M \neq \emptyset.$$

Let us denote $A = \partial k(X,r) \cap \partial M$. Since $A \subset \partial k(X,r)$, all points in A have the same distance from X and this distance is equal to r, $\varrho(X,A) = r$. It remains to be proved that $R = \varrho(X, \partial M)$ is equal to r. There are three options, $R < r, R > r$, and $R = r$. We will show that the first two options are in contradiction with condition (1). The option that $R < r$ is in contradiction with the first part of condition (1), whereas the option $R > r$ is in contradiction with its second part. Only the option $R = r$ complies with both parts of the condition.

Theorem 2. *There is $r_0 > 0$ such that the r-embedding $\widehat{M}_r$ does not exist for $r > r_0$ and $\widehat{M}_r \neq \emptyset$ for $0 < r \leq r_0$.*

Proof. For an arbitrary point $X \in \text{int}\, M$, its distance $r = \varrho(X, \partial M)$ from the boundary ∂M can be calculated. Let us denote $r_0 = \max_{X \in M} \varrho(X, \partial M)$. The compactness of M implies that M bounded, thus there is an upper bound $R_0 \in (0, +\infty)$ for r_0. Then the properties of Euclidean metric imply that there is at least one point $X \in \text{int}\, M$ such that $\overline{k}(X, r_0) \subset M$ and Equation (1) holds. Thus there is an r-embedding $\widehat{M}_r$ for $r = r_0$. For $r < r_0$ we can always find a point X with $\varrho(X, \partial M) = r$ and we can set up a circle $k(X,r)$ in this point.

Let us assume that there is an $r > r_0$ complying with condition (1). Then there must exist a point $X \in M$ such that $r = \varrho(X, \partial M)$. Since we assume $r > r_0$, it should hold $\varrho(X, \partial M) > \max_{X \in M} \varrho(X, \partial M)$, which is a contradiction, therefore such an r-embedding does not exist.

Note 2. The value r_0 is called in the text below the *maximal radius of* $\widehat{M}_r$. There must be at least one point $X \in \text{int}\, M$ whose distance from ∂M is the maximal as in Figure 4.

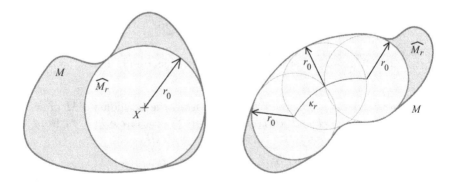

Fig. 4. Examples of r-embedding with maximal radius r_0 indicated

Theorem 3. *Let $r_0 > 0$ be the maximal radius of $\widehat{M_r}$. Then it holds*

$$\operatorname{int} M \subset \bigcup_{r \in (0, r_0)} \widehat{M_r}.$$

Proof. For every point $X \in \operatorname{int} M$ it is possible to find the value of $r = \varrho(X, \partial M)$ as the distance of point X from the boundary ∂M. Therefore, according to Theorem 1 it is possible to construct an open ball $k(X, r)$ complying with condition (1), which means that $X \in \widehat{M_r}$.

Theorem 4. *For every $X \in \operatorname{int} M$ there is exactly one curve parallel with ∂M, κ_r, through X for $r > 0$.*

Proof. The theorem is proved by contradiction: Let us assume that there are two such curves κ_{r_1} and κ_{r_2} through point $X \in M$ with $r_1 \neq r_2$. This means that point $X \in \kappa_{r_1} \wedge X \in \kappa_{r_2}$ and it must hold that $r_1 = \varrho(X, \partial M) \wedge r_2 = \varrho(X, \partial M) => r_1 = r_2$.

Theorem 5. *The boundary ∂M is of M is a special case of κ_r for $r = 0$, i.e. $\partial M = \kappa_0$.*

Proof. Due to the compactness of M it holds $\partial M = \{X \in M, \varrho(X, \partial M) = 0\} = \kappa_0$.

Theorem 6. *Let $r_0 > 0$ be the maximal radius of $\widehat{M_r}$ and let $\{\kappa_r\}_{r \in (0, r_0)}$ be its system of curves parallel with ∂M. Then*

$$\bigcup_{r \in (0, r_0)} \kappa_r = \operatorname{int} M.$$

Proof. The connectedness of set M implies that for an arbitrary $r \in (0, r_0)$ there is at least one point $X \in \operatorname{int} M$ such that $r = \varrho(X, \partial M)$. According to Theorem 4, there is exactly one curve κ_r, where $X \in \kappa_r$.

Theorem 7. *Let ∂M be a smooth curve, $Y \in \partial M$ a boundary point of M. Then $Y \in \widehat{M_r}$ for a suitable $r > 0$.*

Proof. In every point $Y \in \partial M$ it is possible to construct a tangent t to the smooth curve ∂M. Then it is possible to make a circle $k(X, r)$ with a suitable radius $r > 0$ and center $X \in \operatorname{int} M$ so that point Y is also the tangent point of the circle k with tangent t and the closed ball $\overline{k}(X, r)$ complies with condition (1). Since $Y \in \overline{k}(X, r)$, it also belongs to the r-embedding $\widehat{M_r}$.

Theorem 8. *Let $Y \in \partial M$ belong to an r-embedding $\widehat{M_{r_m}}$, for $r_m > 0$. Then Y belongs to infinitely many r-embeddings $\widehat{M_r}$ for $r \in (0, r_m)$.*

Proof. If $Y \in \partial M$ belongs to a certain r-embedding $\widehat{M_{r_m}}$, then there is a closed ball $\overline{k}(X, r_m)$ such that $Y \in \overline{k}(X, r_m)$ and $k(X, r_m) \subset \operatorname{int} M$. From the geometry of a circle it is evident that it is possible to inscribe an arbitrary circle $k(X', r)$

into $k(X, r_m)$ such that $r \in (0, r_m)$ with the properties $Y \in \partial k(X', r)$ and $k(X', r) \subset \mathrm{int}\, M$. Thus $k(X', r)$ fulfills condition (1) and point $Y \in \widehat{M}_r$. There are infinitely many coverings of this kind because r can be chosen arbitrarily in the interval $(0, r_m)$.

Note 3. It is worth pointing out that r_m in the previous theorem is less or equal to the radius of curvature of the boundary ∂M.

Theorem 9. *Let ∂M be a smooth curve and let $r_0 > 0$ be the maximal radius of $\widehat{M}_r$. Then it holds*

$$M = \bigcup_{r \in (0, r_0)} \widehat{M}_r. \tag{2}$$

Proof. According to Theorem 3, $\mathrm{int}\, M \subset \bigcup_{r \in (0, r_0)} \widehat{M}_r$ and according to Theorem 7 every point $Y \in \partial M$ belongs to a certain r-embedding, therefore also the boundary $\partial M \subset \bigcup_{r \in (0, r_0)} \widehat{M}_r$. Since $\mathrm{int}\, M \cup \partial M = M$, we can say that $M \subset \bigcup_{r \in (0, r_0)} \widehat{M}_r$. Definition 1 also says that $\bigcup_{r \in (0, r_0)} \widehat{M}_r \subset M$, therefore there must be an equality between the sets in Equation (2).

3.2 The Discrete Description

So far we were dealing with the theoretical model in $\mathbb{E}^2$. In the real application we have to deal with the discrete problem, where the set M is discretized to pixels. The boundary ∂M is represented with boundary pixels. The set M will be now called the *object O*, it is a connected (better 4-connected) set of pixels which may be obtained by thresholding or other methods of image segmentation. Its boundary will be denoted by ∂O. A curve parallel with the object boundary will be denoted with $\mathcal{K}_r$ and the r-embedding $\widehat{M}_r$ will be now $\widehat{O}_r$. The transition from $\mathbb{E}^2$ is ambiguous, it depends on the chosen type of connectivity of curves. It is complicated to rephrase and prove all the theorems from $\mathbb{E}^2$ in the discrete case. The algorithm for the discrete case can be considered as a numerical method for solving the problem.

For the recrystallization, we will need the distance of each point of object O from the object boundary ∂O. As Theorem 4 says for the continuous problem, there is exactly one equidistant through each point of the set. Finding these curves is a way to estimate the distance of each pixel from the object boundary. The method of step-by-step erosion of the object is used for this purpose. It means that in each step we check all pixels of the object and if there is a background pixel in the neighborhood $N_4(p)$ or $N_8(p)$ of a pixel $p \in O$, then this pixel is changed to a background pixel (For more information about the N_4 and N_8 neighborhoods see [3]).

In each step we assign to each pixel the step number of the step in which the pixel was eroded. The step numbering starts from zero so that the boundary pixels are assigned zero. The erosion process gives us an estimation of parallel curves. The use of $N_4(p)$ neighborhoods for the erosion implies that the parallel

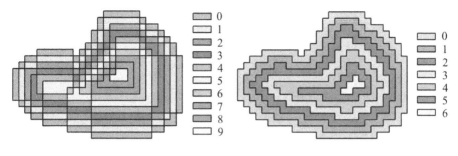

(a) 4-connected curves $\mathcal{K}_r$ parallel with the boundary of object O

(b) 8-connected curves $\mathcal{K}_r$ parallel with the boundary of object O

Fig. 5.

curves are 4-connected, whereas $N_8(p)$ neighborhood give us 8-connected curves, as it is illustrated in Figure 5.

Now we start constructing the r-embedding $\widehat{O}_r$ of object O from the furthest curve $\mathcal{K}_r$, where $r = r_0$. The curve $\mathcal{K}_r$ need not be connected, it may be composed of more connected components $\mathcal{K}_r^i$. We inscribe circles to the object with centers in the pixels that belong to the curves $\mathcal{K}_r^i$ and the union of these circles is made. Thus we obtain sets of pixels O_r^i, which need not be disjoint. In the case when object O_r^i is disjoint from the rest of the objects, this set will be considered as a new nucleus C_j, where j is the index of the nucleus. If the set O_r^i has a non-empty intersection with object O_r^l, where $i \neq l$, we have to decide if there is one nucleus or two nuclei with an intersection. Both situations are shown in Figure 6. If the curve $\mathcal{K}_r^i$ has a non-empty intersection with object O_r^l, where $i \neq l$, there is one nucleus, in the other case there are two of them.

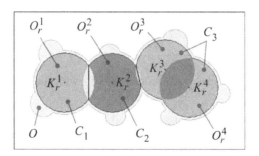

Fig. 6. Initial nuclei of elementary objects

Let us now focus on step-by-step enlarging the nucleus and thus creating of more nuclei. The situation with one initial nucleus is illustrated in Figure 7.

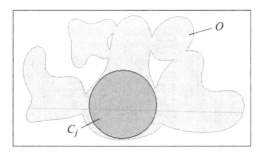

Fig. 7. Initial nucleus of the object

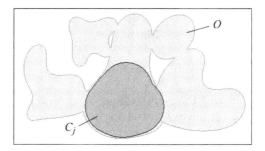

Fig. 8. Growth of the object nucleus

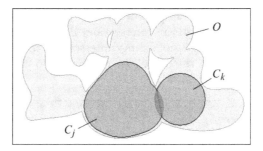

Fig. 9. Construction of a new nucleus with intersection

In the iterative steps that follow, the value of r is decreased by one in each step to create the next r-embedding $\widehat{O}_r$. This r-embedding may be composed of more sets O_r^i, which correspond to curves $\mathcal{K}_r^i$. If a curve parallel with the boundary of the object has a non-empty intersection with a nucleus C_j, this nucleus is enlarged by that curve as it is illustrated in Figure 8. In the opposite case a new nucleus is created. If some nuclei have a non-empty intersection after being created or enlarged, the elements (pixels) of the intersection are assigned to nuclei whose disjoint parts (nucleus minus the intersection) are closer to that pixels, as it is illustrated in Figure 9. The whole process is iterated while $r > 0$.

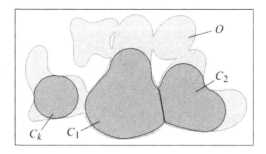

Fig. 10. Creating a new nucleus with no intersection

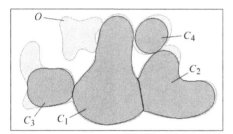

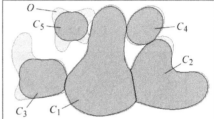

(a) Gradual growth of nuclei and formation of the fourth nucleus

(b) Gradual growth of nuclei and formation of the fifth nucleus

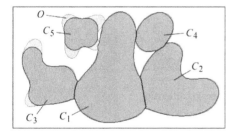

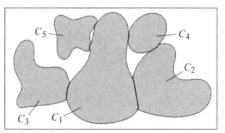

(c) Gradual growth of all nuclei

(d) Final partition

Fig. 11.

After the process is terminated, we have obtained a partition of the object O in separate disjoint objects C_j.

The step-by-step growth of separate nuclei form the final partition of the object is illustrated in Figure 11.

In the partition of the whole object it is noticeable that the nuclei that were formed earlier overlap slightly with other nuclei, but this discrepancy is negligible as far as the number of elementary objects and the estimation of their size are concerned.

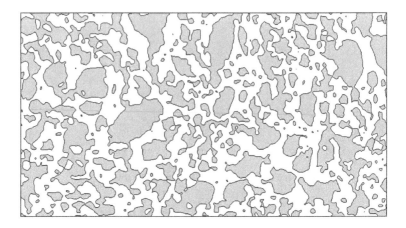

Fig. 12. Objects before recrystallization

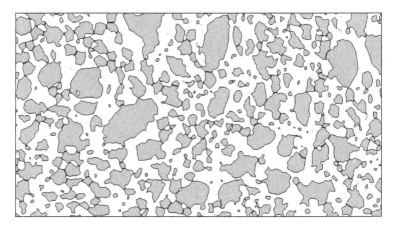

Fig. 13. Objects after recrystallization

In the case when the object is disjoint with all existing nuclei C_i, a new nucleus C_k is created. This is illustrated in Figure 10.

The method of recrystallization has been used in a practical application. Figure 12 shows objects after the recrystallization and Figure 13 after recrystallization.

4   Conclusion

In this contribution, we presented a new method for splitting bunches of elementary objects detected by image segmentation as one object into more elementary objects. The method was software-implemented and has been used for analysis of images in food industry in collaboration with the University of Veterinary and Pharmaceutical Sciences Brno. It has a general application e.g. in metallurgy, medicine, and all applications of image processing where touching or slightly overlapping objects occur. The next step after having the object split in several objects is reconstruction of the original shape of the elementary objects, which is what we focus on in our further research.

Acknowledgement. This work was supported by grant FSI-J-13-1997 of the Faculty of Mechanical Engineering, Brno University of Technology. Publication of the results was financially supported by the project Popularization of BUT R&D results and support systematic collaboration with Czech students CZ.1.07/2.3.00/35.0004.

References

1. Bai, X., Sun, C., Zhou, F.: Splitting touching cells based on concave points and ellipse fitting. Pattern Recognit. 42, 2434–2446 (2009)
2. Cheng, J., Rajapakse, J.C.: Segmentation of clustered nuclei with shape markers and marking function. IEEE Trans. Biomed. Eng. 56, 741–748 (2009)
3. Herman, G.T.: Geometry of Digital Spaces. Springer (1998)
4. Kumar, S., Ong, S.H., Ranganath, S., Ong, T.C., Chew, F.T.: A rule-based approach for robust clump splitting. Pattern Recognit. 39, 1088–1098 (2006)
5. Martišek, D.: Matematické principy grafických systémů. PhDr. Karel Kovařík, nakladatelství Littera. Brno, Czech Republic (2002) ISBN 80-857-6319-2
6. Pham, D.L., Xu, C., Prince, J.: Current Methods in Medical Image Segmentation. Annual Review of Biomedical Engineering 2, 315–337 (2000)
7. Pratt, W.K.: Digital image processing: PIKS inside, 3rd edn. Wiley, New York (2001)
8. Štarha, P., Druckmüllerová, H., Tremlová, B.: Numerical Analysis of Color Hue Component in Digital Images. In: Mendel 2011, 17th International Conference of Soft Computing, pp. 497–503, Brno University of Technology, Brno (2011)
9. Svensson, S.: Aspects on the reverse fuzzy distance transform. Pattern Recognit. Lett. 29, 888–896 (2008)
10. Vincent, L., Soille, P.: Watersheds in digital spaces: an efficient algorithm based on immersion simulations. IEEE Trans. Pattern Anal. Mach. Intell. 13, 583–598 (1991)

Calibrationless Sensor Fusion Using Linear Optimization for Depth Matching

László Havasi[1], Attila Kiss[1,2], László Spórás[1,3], and Tamás Szirányi[1,3]

[1] Distributed Events Analysis Research Laboratory, Institute for Computer Science and Control, Hungarian Academy of Sciences, Budapest, Hungary
[2] Department of Computer Science of L. Eötvös University, Faculty of Science, Budapest, Hungary
[3] Peter Pazmany Catholic University, Budapest, Hungary
{laszlo.havasi,attila.kiss,tamas.sziranyi}@sztaki.mta.hu

Abstract. Recently the observation of surveillanced areas scanned by multi-camera systems is getting more and more popular. The newly developed sensors give new opportunities for exploiting novel features.

Using the information gained from a conventional camera we have data about the colours, the shape of objects and the micro-structures; and we have additional information while using thermal camera in the darkness. A camera with depth sensor can find the motion and the position of an object in space even in the case when conventional cameras are unusable.

How can we register the corresponding elements on different pictures? There are numerous approaches to the solution of this problem. One of the most used solutions is that the registration is based on the motion. In this method it is not necessary to look for the main features on the pictures to register the related objects, since the features would be different because of the different properties of the cameras. It is easier and faster if the registration is based on the motion. But other problems will arise in this case: shadows or shiny specular surfaces cause problems at the motion.

This paper is about how we can register the corresponding elements in a multi-camera system, and how we can find a homography between the image planes in real time, so that we can register a moving object in the images of different cameras based on the depth information.

Keywords: Motion detection, Multi-camera system, Image matching, Homography.

1 Introduction

In crowded places it is difficult to track correctly individual people from a single point of view with a conventional camera. That is why we started to investigate multi-camera systems for a better tracking algorithm. We try to fit different camera images onto one common image plane in order to gain a 3-dimensional picture of the investigated territory in real time. Plenty of research results exist in the literature in connection with this topic when one tries image registration

R.P. Barneva, V.E. Brimkov, and J. Šlapal (Eds.): IWCIA 2014, LNCS 8466, pp. 158–170, 2014.

or fitting objects. It's more difficult to distinguish different objects on an image because of the occlusion. If we have more images from different angles about the same screen then we will have more information about that fixed place and the objects in it. By setting an image as the reference image, we can identify the ground plane on every image. Hence we can match the ground points of the objects. If we would shift the ground plane in the direction of its normal vector then we would cut the objects in the same position in all the images. Furthermore we could get more information from a view from above: we could get the spatial position of each object, if we could register the images in different planes together. This plane might be shifted to scan the intersections in the common space of the views [10].

A common plane of all views has an outstanding role: [12] assumes that a reference plane is visible in all views, and shows for this case that lines and cameras, as well as, planes and cameras have a linear relationship. Consequently, all 3D features and all cameras can be reconstructed simultaneously from a single linear system, which handles missing image measurements naturally.

We used the idea of Khan et al. [10] and developed a method that applies planar homographic occupancy. The present method is developed for multiple planes parallel with the center line of a depth sensor applied together with at least one of the cameras (this is different from the method of Khan and Shah). We try to fit the different image planes onto one screen and try to find different homographies for this, in order to shift these planes into different depth values, then getting an exact match of corresponding moving objects on different image planes. We use a combinatorial algorithm to find these homographies applying a linear programming algorithm from the literature of combinatorial optimization. This algorithm is developed with applying an optimization method based on the well-known golden ratio [9].

The aforementioned method can be used if we have more cameras than two or three, for example on a football match we can follow the movement of the players with this algorithm. In practice there are few events where multiple-camera systems are necessary. In these cases there are different methods for point to point registration and we chose the most efficient one according to our problem. Camera observation is usually integrated into busy places hence there exists motion in all the images. We can make motion mask statistics about these movements, related to a reference image. By detecting motion on the pixel array of this image we make a conditional statistics to the motion masks of the reference image and to all the other images. We will have different motion statistics for each pixel of each reference image belonging to different cameras exactly as many statistics as many cameras we used to integrate into the system [15]. (See Fig. 1.) The structure of this paper is first we introduce an algorithm for the image matching. In particular we will talk about the motion detection, the fitting of the image planes and the mathematical problems that we found and the mathematical background of our solution. In the next section we will mention some experimental results. And we will mention some future directions of this research, some possible ways to carry on with this work.

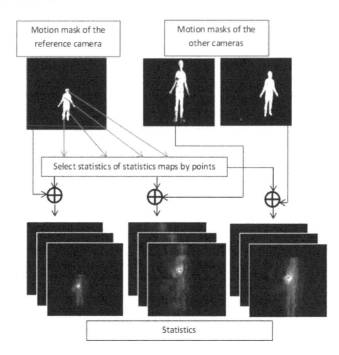

Fig. 1. Images of the first row are the motion masks. We build the motion statistics maps, assigning places where motion exists (white pixel) on the reference motion mask. Then we add all motion masks to statistics belonging to it, including the reference motion mask. If we have enough number of measurements, we can pair correlated moving pixels related to the reference camera. This conditional statistics of motion map consists of Gaussian distributions. The most probable pair of a point of the probabilistic map is that one of the highest probability. In the probabilistic pairing we assign not only one point but a small neighborhood around it. The red areas of the statistics represent the neighborhood around the maximum.

2 Image Matching

The main steps of the algorithm for image matching based on the depth values are as follows:

1. Own motion detection algorithm for the depth sensor.
2. From the motion masks, we can get the coordinates of different points on all images, and we will show how it becomes possible to register the corresponding points.
3. What about the homography? Different problems are coming up if these corresponding points are belonging to planes that are orthogonal to the ground plane and have different depth values. How can we deal with these problems with a depth sensor?
4. Now we have a homography to project corresponding points from the same plane onto a different plane. We give a method to modify this homography in

order to calculate different homographies into parallel planes to the original one.

5. We need to define the variable γ that doesn't change in a linear manner while the change of the depth values is linear.

2.1 Depth Motion Detection

For the cameras without depth sensors we use the built in class of OpenCV. This class is BackgroundSubtractorMOG. Unfortunately this motion detector does not give perfect results, furthermore the output will be too noisy. If we have more moving objects in occlusion with each other, then we are not able to find their original depth values. That was the motivation why we developed an algorithm for depth sensors, that determines the motion from the change of depth values [13]. So we get more reliable and less noisy motion masks from the depth sensor and these are important for further measurements.

The fundamental idea is that the sensor collects data in the first step, and we store them. If we have collected enough data, the algorithm can create an initial background mask from this database. Each pixel of this initial mask will get the average value of the measurements in that pixel. If there is a moving object, then the depth values of this object will be smaller, then the values of the background mask, and the depth values of the moving object will be added to the foreground mask. The algorithm can learn, for example if an object in the foreground mask doesn't move, and the depth values of it are not changing, and these values are lower than a threshold, then the background mask will be updated with these values. Furthermore if the depth values of an object suddenly get larger (after an object has moved away the depth values of the background became larger), then the algorithm updates the average of the depth values to the background mask [2,6,3]. (See Fig. 2.)

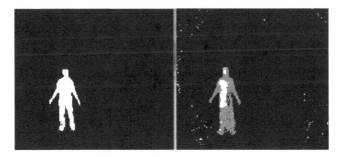

Fig. 2. The image on the right was made with the in-built motion detector class of the OpenCV. We can see, that the motion mask is not perfect, and the image contains noisy pixels. The image on the left has a good motion mask without noise as the result of the proposed method.

2.2   Corresponding Motion Statistics

It is necessary to find accurate point pairs of high correspondence in order to obtain a good homography between two image planes. The size and the shape of a moving object may be significantly different in images of different cameras, and matching corresponding points is more difficult and becomes less accurate if we work with more cameras. That was the reason why we create co-motion statistics for each camera. We applied the idea of investigating corresponding motions from a previous work of the authors László Havasi, Tamás Szirányi and our former colleague and co-atuhor Zoltán Szlávik [15]. We need a reference camera, fixing its image plane as the reference plane where we will fit the image planes of the other cameras with the proper homographies. This camera must need a depth sensor, so we used a professional camera with depth sensor, namely Mesa Imaging Swiss Ranger 4000.

If our algorithm detects motion in any pixel of the reference image, then we update the statistics of this pixel position for each camera with the current motion mask [15].

For memory and run-time reasons we re-size the reference image and motion maps to size 80 * 80. For example we set a motion statistics map with 80 * 80 size to each camera, and each element of that map is a statistics with size 80 * 80. Let us suppose that we sense motion in the coordinates (45, 35) of the reference image, then we add the motion mask of the related image to the statistical map of the pixel in the coordinates (45,35) of the reference camera. Trivially we have to re-size the motion masks of the other cameras to size 80*80 too. Experimental results showed that the resizing for this size keeps the sufficient information for the given task.

2.3   Problem with the Homography

The cameras are sensing the world from different point of views, so the corresponding objects will have different sizes on their images. We used homography to fit the different images onto a reference image. We need four point pairs to find a homography between two image planes. A point pair is an element from the Cartesian product of the set of the vertices of the reference image plane and another image plane. We are looking for corresponding points on these image planes, but after the fitting the transformation may deform the picture (because the image planes are not have to be coplanar), moreover, the transformation may rotate the whole image plane.

But if the corresponding points would lie in the same plane it would not be sufficient neither. Because the fitting would be accurate only in their plane, but it would be inaccurate in all the other parallel planes [4,7,14]. (See Fig. 3.)

Our task is to project these point pairs onto the same image plane and make a procedure that can shift this plane into different depth values. When this procedure is fast enough that will make us able to match the corresponding moving objects on different camera images in real time.

In order to ensure that the corresponding points lie in the same plane, we use a depth filter on the reference camera to choose that plane on which we want to

Fig. 3. In these pictures one can see in the first column the outputs of a depth sensor, in the second column the outputs of a conventional camera and in the third column the image after merging the motion masks of these two sensors. First row: The object is placed behind the proper plane of the homography. Second row: The object is placed in the proper plane of the homography. Third row: The object is placed before the proper plane of the homography.

find a homography. Now the reference camera will perceive the motions only in the environment of the selected depth, hence it will generate motion masks only in that depth. This will result motion statistics that we needed to have for our algorithm. The point pairs that can be achieved from these statistics will be located on the same plane. (See Fig. 4.) For further details the reader is referred to [14].

2.4 Pushing the Plane of the Homography Depending on Depth

After we have found a good homography for the points in the same depth, we only have to solve the parallel shifting of homography calculation for different source planes.

We used the fusion of multiple planes to increase robustness and accuracy of our method. Our method performs fusion of different image planes onto one reference plane. One can see that conventional feature correspondence-based methods are not feasible for homography calculus among image planes. For example if we have a homography $H_{p,q}$ induced by a reference plane r between to different views p and q then the homography $H_{p,q}$ induced by a plane s parallel to r is given by the following formula: (for further details the reader is referred to [10]:

$$H_{psq} = (H_{prq} + [0|\gamma v_{ref}])(I_{3x3} - \frac{1}{1+\gamma}[0|\gamma v_{ref}]) \tag{1}$$

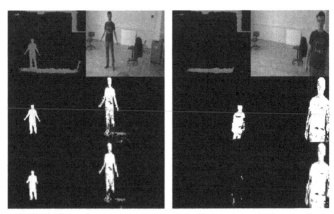

(a) Object in the plane of the (b) Object in a different plane
homography

Fig. 4. The depth filter: actual distance is: 3.5m from the camera with 0.3m depth range. Valid geometrical depth is 3.2 – 3.8m. In the figures in the first column there is an image from the depth sensor when we apply a depth filter on it and the motion mask of the depth sensor without the depth filter and the intersection of the two images in the first two rows, from top to bottom respectively. In the second column there is an image from a conventional camera and the motion mask of that image and a eroded shrunk image from the motion mask of the conventional camera, from top to bottom respectively. The bottommost pictures are applied during the construction of the statistics maps of the co-motion.

Where v_{ref} is the vanishing point of the normal direction and γ is a scalar multiple that has to control the distance between the parallel planes. Typically we use the plane orthogonal to the center line of the depth sensor of the reference camera as the reference plane and the parallel direction with that line as the reference direction. Here the homographies were determined with SIFT [11] feature matches and using the RANSAC algorithm [8].

2.5 Defining the γ Parameter

If the value γ approaches the $\pm\infty$, then the homography H_{psq} converges to stable states. Experimental results showed that the final value of H_{psq} is well approximated on the interval [-20, 20]. (See Fig. 5.)

After we detected a homography we wanted to find a method to refresh the optimal value of the scalar multiple γ for the best fitting if we shift our reference plane into different depth. We applied a combinatorial method to find this optimal γ. First we calculated a range small enough to start the search in with random evaluating an error function indicating the value of correct fitting. After that we applied a convex optimization method, a specially modified version of the well-known interval halfing method, applying the golden ratio values for the shrinking of the interval. We could do this because we managed to prove that

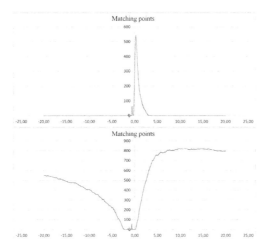

Fig. 5. We used two different motion masks. One is the motion mask of the reference camera the other one is the motion mask of another camera that we want to fit. The first graph is about the case when the matching function is calculated in an image plane with larger depth value than the the reference plane has. The other graph is about a matching function in the reference plane. The axis X shows the value of γ, the axis Y is the amount of matching points after the fitting. A coordinate on the motion mask of the reference image will be matched if our algorithm senses motion in that coordinate and in that pixel there is motion in the motion mask of the other camera too. Our algorithm is looking for a fitting that maximizes the amount of the matched points between two images.

this error function using this value γ is convex with probability 1. We made some tests and it resulted that our algorithm runs in real time. In comparison with other methods, there are some algorithms trying to solve problems similar to our, for example in the article [1] we find similar arrangement of equipments, but that solution takes about a minute per frame to get depth information, which is much slower than our present real-time approach.

3 Experimental Results

We used this system for an indoor scene because our camera with the depth sensor is unsuitable for outdoor recording, the algorithm itself is applicable in outdoor territories too. However the base distance of cameras is a main parameter of our algorithm. If this distance is getting larger then the imprecision of the matching will be bigger. Furthermore the response time will be slower, if more cameras are integrated into the system (see Figs. 6 and 7), or we increase the resolution of the cameras.

Currently the system works in case of only one object (see Figs. 9 and 10). However if we can collect enough information with this single object about the γ

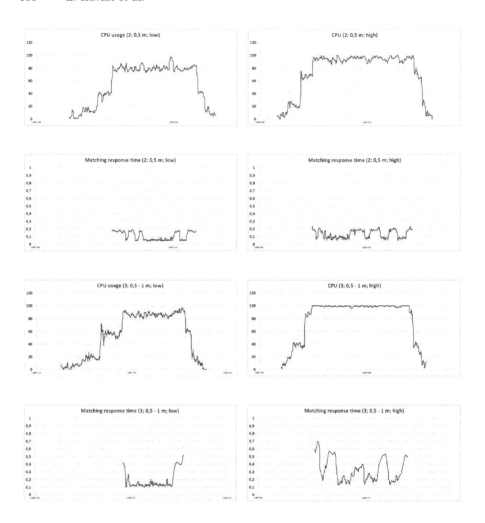

Fig. 6. Figures in the first and third lines show the CPU usage where the axis x is the time (CPU time) and the axis y is the CPU utilization in percentage. We started the measurements from the idle state of the CPU. The steps of each measurement were the following: start the sampling of the cameras one by one, start the generating of the motion masks, start the fitting phase (two different types of result, successful or unsuccessful), stop the fitting phase, stop the generation of the motion masks, stop the sampling of the cameras. Figures in the second and the fourth lines show the response time of the algorithm where the axis x is the time (CPU time) and the axis y is the response time in second. The first set of figures of the first two rows are belong together and their axis x is uniformly scaled in order to get a better view. The different parametrization of measurements are indicated in the title of the figure in the form (n, d, r) where n stands for the number of cameras we used, d stands for the distance of the cameras from the reference camera and r stands for the resolution we applied (high or low). The resolution of the Mesa camera is (176x144), the two different resolutions we used for the Axis Network camera are (320x180) and (800x450) and for the Axis Thermal Network camera are (240x180) and (480x360).

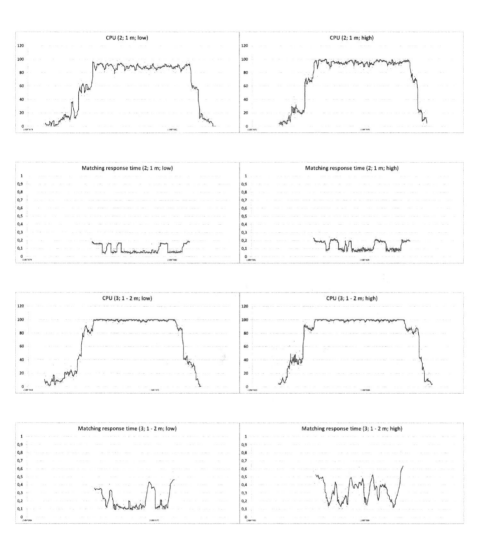

Fig. 7. These figures show what happens if we increase the distances between the cameras and the reference camera

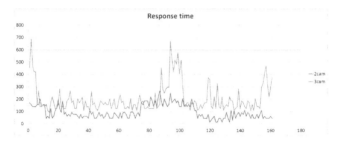

Fig. 8. Response time of the matching with 2 or 3 cameras. Red line is the measurement of 2 cameras, blue line is the measurement of 3 cameras. Axis X is the serial number of frames, axis Y is the running time in millisec. If the algorithm finds a good matching, the response time will be under 100 millisec in case of 2 camera (between 30 - 70), if do not find, then this time will be larger (between 80 - 100). In the other case these time values are higher.

Fig. 9. The result of 2 cameras. We used a depth sensor (Mesa Imaging Swiss Ranger 4000) and a visible spectrum camera (Axis Optical Camera). The first two images are the raw data from cameras in the first row. The first image of the second row is the matching with a plane gained by pushing the plane of the homography, second image is the original plane of the homography without pushing. Third image of the first row is the result on the raw data with the matching homography.

values of different depth, then the matching algorithm can be omitted, simply we can search in a database to find the value γ of the current depth. This method eliminates the need for a matching algorithm handling more objects. Since we have a database of pairs of depth and γ values, it is possible to estimate from the picture of depth camera that one or more objects moving in front of the cameras. After the database has been built up, we are left only a few possibilities to choose the proper depth value and the object helps us to determine the exact depth value because the matching of the objects would not be possible in another plane with the given value γ.

Fig. 10. The result of 3 cameras. We used a depth sensor (Mesa Imaging Swiss Ranger 4000), a visible spectrum camera (Axis Optical Camera), and a thermal camera (Axis Thermal Camera). The first three images are the raw data from cameras. The first image of the second row is the matching with a plane gained by pushing the plane of the homographies, second image is the original plane of the homographies without pushing. Third image of the second row is the result on the raw data after the matching with the homographies.

We made a figure to compare the response times of the different cases. (See Fig. 8.)

4   Conclusions

The paper presents a new method for camera fusion. We used co-motion statistics to register the image planes together, and a combinatorial algorithm to find depth values to each image planes. Our idea was to combine these methods and to use the golden ratio in the convex optimization step. It may provide a solution to application cases like safety systems operating with more cameras. It also helps to make a conclusion in case of the breakdown of a camera, caused that by an intruder target person, if the rest of the cameras perceive him. If the system has already finished the learning phase, then the remaining cameras are enough to recognize the target person on their images and determine the position of the intruder target person in the image of the camera that was broke down.

Acknowledgments. This work has been supported by the European Community's Seventh Framework Programme (FP7-SEC-2011-1) under grant agreement n. 285320 (PROACTIVE project). The research was also partially supported by the Hungarian Scientific Research Fund (grants No. OTKA 108947 and OTKA 106374).

References

1. van Baar, J., Beardsley, P., Pollefeys, M., Gross, M.: Sensor fusion for depth estimation, including TOF and thermal sensors. In: IEEE Second International Conference on 3D Imaging, Modeling, Processing, Visualization and Transmission (3DIMPVT), pp. 472–478 (2012)
2. Barnich, O., Van Droogenbroeck, M.: ViBe: A universal background subtraction algorithm for video sequences. IEEE Transactions on Image Processing 20(6), 1709–1724 (2011)
3. Barnich, O., Van Droogenbroeck, M.: ViBe: a powerful random technique to estimate the background in video sequences. In: International Conference on Acoustics, Speech, and Signal Processing (ICASSP 2009), pp. 945–948 (2009)
4. Coxeter, H.S.M.: Projective geometry. Springer (2003)
5. Criminisi, A., Reid, I., Zisserman, A.: Single view metrology. International Journal of Computer Vision 40(2), 123–148 (2000)
6. Van Droogenbroeck, M., Paquot, O.: Background subtraction: Experiments and improvements for ViBe. In: Change Detection Workshop (CDW), Providence, Rhode Island, pp. 1709–1724 (2012)
7. Hartley, R.I.: Theory and practice of projective rectification. International Journal of Computer Vision 35(2), 115–127 (1999)
8. Hartley, R., Zisserman, A.: Multiple View Geometry in Computer Vision. Cambridge University Press, Cambridge (2002)
9. Illés, T.: Nonlinear Optimization. University Lecture Notes, L. Eötvös University of Science, Budapest, Hungary (in Hungarian)
10. Khan, S.M., Shah, M.: Tracking multiple occluding people by localizing on multiple scene planes. IEEE Transactions on Pattern Analysis and Machine Intelligence 31(3), 505–519 (2009)
11. Lowe, D.: Distinctive image features from Scale invariant keypoints. International Journal of Computer Vision, 91–110
12. Rother, C.: Linear multiview reconstruction of points, lines, planes and cameras using a reference plane. In: Proc. Ninth IEEE International Conference on Computer Vision, vol. 2, pp. 1210–1217 (2003)
13. Shotton, J., et al.: Efficient human pose estimation from single depth images. In: Decision Forests for Computer Vision and Medical Image Analysis, pp. 175–192 (2013)
14. Spórás, L.: Sensor fusion of depth sensors, color and thermal cameras with small base distance (Bachelor's Thesis), Peter Pazmany Chatolic University, Budapest, Hungary (2013) (in Hungarian)
15. Szlávik, Z., Szirányi, T., Havasi, L.: Stochastic view registration of overlapping cameras based on arbitrary motion. IEEE Tr. Image Processing 16(3), 710–720 (2007)

Optimal RGB Light-Mixing
for Image Acquisition Using Random Search
and Robust Parameter Design

HyungTae Kim[1], Kyeongyong Cho[2], SeungTaek Kim[1], and Jongseok Kim[1]

[1] Manufacturing System R&D Group, KITECH, 35-3, HongCheon,
IpJang, CheonAn, ChungNam, 331-825, South Korea
{htkim,stkim,jongseok}@kitech.re.kr
[2] UTRC, KAIST, DaeJeon, 305-701, South Korea
yong00@kaist.ac.kr

Abstract. Obtaining a fine image is one of the major issues in industrial vision, and light mixing techniques are one of the alternatives. Auto-lighting using a multiple color mixer requires iterative actions. Random search shows high efficiency in finding the optimal illumination. However, random search is one of the numerical algorithms to find local minimum, so the algorithm parameters affect the performance of auto-lighting. The relation between the light mixing and the image fineness is mathematically nonlinear, and it is difficult to tune the parameters reliably. This study proposes a method to determine reliable parameters in random search for optimal illumination in image inspection using a color mixer. The Taguchi method was applied to maximize the image fineness and minimize iterations. The parameters selected for Taguchi analysis were the initial voltage, initial variance, and convergence constant. An $L_{25}(5^5)$ orthogonal array was constructed in consideration of the 5 parameters and 5 levels. The determined parameters were applied to retests, which showed fewer iterations and the acquired image was close to the best case.

Keywords: Optimal illumination, Light control, Image sharpness, Maximum distinctness, Taguchi method, Experiment design.

1 Introduction

Focus, illumination, and exposure are the major environmental factors to acquire fine images in industrial machine vision. These environmental factors are adjusted to obtain fine images after fixing a target object. It is time-consuming work to finding the optimized environmental factors, so automatic systems are necessary. Automatic focusing is widely used in industrial machine vision and focusing methods have been reported in many studies[9], [2], [12]. Automatic focusing is a simple process to maximize image sharpness by adjusting and moving optical units attached in front of a camera. The image sharpness is a kind of contrast and degree of focus calculated from pixel-based operations. Mathematical methods have been proposed for the sharpness, and their concept usually

R.P. Barneva, V.E. Brimkov, and J. Šlapal (Eds.): IWCIA 2014, LNCS 8466, pp. 171–185, 2014.
© Springer International Publishing Switzerland 2014

involves computing differences of pixels in an image[26],[17],[28],[37],[3],[36]. Although illumination is also one of the important environmental factors to obtain fine images, it is not noticed much in machine vision. Automatic lighting is partially applied for a single optical source, but there have been relatively few related cases and they does not provide an efficient solution to optimize multiple color-mixing sources, so trial-and-error is currently the only way to find optimal light conditions.

The sharpness is the successive response from the optical spectrum of illumination, the reflection of a target object, the optical system, and the camera. So, images acquired in industrial machine vision can be improved by adjusting the spectrum and intensity of illumination, which is called mixed color vision. Although the procedure from illumination to sharpness can be written by mathematical equations, it is actually difficult to simulate them due to physical properties and boundary conditions[1]. Mixed color techniques were recently applied to automated optical inspection (AOI). There is demand for automatic mixed color vision, but there are not many practical solutions. The topics of the mixed color vision are divided into passive and active color vision. Passive color vision uses a wide-range optical source or a white source, with various color filters and cameras. A target object reflects wide-range spectral light, and the image quality is affected by specified bands and the intensity of the spectrum. One pair of a band filter and a camera acquires a specified spectral image, and a new image is constructed from the synthesis of multiple pairs. The synthesis involves pixel-based operations to create a fine image by combining multiple monochrome images acquired in various optical bands. The representative pixel-based operations are simple averaging, weighted averaging, and geometric averaging in color coordinates. Passive color vision is applied to AOI systems for flat-panel display (FPD) manufacturing[35]. However, manual trials and rules-of-thumb are applied to set the combination of pairs and the image synthesis. It is laborious and time consuming work to find the optimal combination and operation. This approach is also used for agricultural products, such as tomatoes [8], potatoes[25], cucumbers[7] and seedlings[31].

Active color vision acquires an image under a mixed light from multiple color sources which are optimized for image quality. Active color vision systems must include an optical system, called a color mixer, to mix multiple color sources. A target object is placed under the color mixer and a monochrome camera acquires an image of the current lighting conditions. Color mixing mechanisms have been proposed in many studies. The mixing principle, uniformity, and color accuracy are major issues in this area. The light engine[24], tunable lamp[30], light pipe [6], micro-lens array[23] and collimator[32] are the representative techniques for color mixing. The performance of these techniques was evaluated at the optical physics level, so the applications to machine vision has not been noticed much. Therefore, we have been considering the usage of color mixers in view of machine vision with focus on image quality. The sharpness commonly used in auto-focus can be applied to check if the current illumination status is suitable for image acquisition. As for auto-focus, the optimal illumination in active color vision is a lighting status to maximize the sharpness after adjusting multiple color sources. The optimal illumination can be found with a large amount of iterations using optimum methods and numerical algorithms, so it can be considered as automatic lighting. Most of the auto-light is based on iterative methods, with equal step search, but the iteration can be reduced using single-variable search methods[13]. Methods of non-uniform step size were applied to some reports such as single variable optimum and direct searches[18,11]. Active color vision was recently reported in a few studies. The typical principle of this optimal illumination is to maximize the sharpness by adjusting the inputs of the multiple sources. Vriesenga showed how to select the optimal illumination for FPD inspection, but the approach was experimental with limited results, and it was hard to apply to general cases[33]. Mixed light can be exactly reproduced from a target spectrum optimized for image acquisition and found in preliminary experiments. Multi-spectral approaches were applied to endoscopes and a multi-flash system, but they did not show how to automatically find the optimal light[19,4]. We presented a histogram-based and a mathematical approaches for auto-light using RGB mixing, but it was quite iterative and time-consuming[14]. Auto-light for quick optimal illumination can be conducted using optimum methods to maximize the sharpness within a short time by adjusting the inputs of multiple sources[15]. The auto-light is a search algorithm, so it is important to set the best algorithm parameters to increase search performance. In this study, the Taguchi method is discussed for tuning the random search algorithm to increase searching speed and maximize the image sharpness.

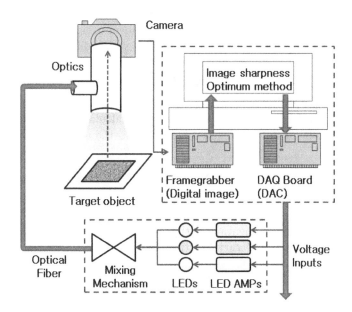

Fig. 1. Schematic diagram of image acquisition with a color mixer

2   Random Search for Optimal Illumination

2.1   Color Mixing and Cost Function

The active color vision system is described in diagram 1. A monochrome camera acquires an image and is connected to a frame grabber through a CAMLINK cable. The framegrabber acquires the image in PC memory. The sharpness and optimum algorithm is loaded in the PC. Voltage inputs for the multiple color sources are determined by the algorithm and generated through DAC ports in a DAQ board. An individual AMP drives current for power LEDs by voltage inputs. The lights from the LEDs are mixed in a mixing mechanism. The mixed light is transferred to optics through optical fiber and used to illuminate a target object. The camera acquires the image under mixed light. The sequence from the image acquisition to illumination forms a feedback loop. The aim of optimal illumination is to acquire fine and clear images by adjusting multiple light sources in a color mixer. This problem is defined to find desirable voltage inputs of the lights in the condition that maximizes image sharpness. So, it is necessary to define equations between the voltage inputs and the sharpness. The equations were proposed in our previous research[15] and are briefly summarized in this chapter. The color mixer has arbitrary N light sources and is driven by respective voltage inputs, so the following vector for voltage inputs can be defined.

$$V = (v_1, v_2, \ldots, v_N) \tag{1}$$

The voltage inputs change light intensity and color, which finally affects the grey level of pixels $I(x, y)$ in an image. This can be shown using a spectral form, but it is replaced with an arbitrary function $h_{xy}(V)$ which differs in every pixel of the image.

$$I(x, y) = h_{xy}(V) \tag{2}$$

There are several definitions proposed for the sharpness, but discrete variance is the most common form. The sharpness σ can be calculated as follows from the grey level of pixels for an $m \times n$ size image[29]. $\bar{I}$ is the average grey level of an image.

$$\sigma^2 = \frac{1}{mn} \sum_{x}^{m} \sum_{y}^{n} \left(I(x, y) - \bar{I} \right) \tag{3}$$

The sharpness can be varied when the mixed light from the color mixer is changed by adjusting the inputs. As mentioned, the mathematical formulation is complex, but we can use a simple arbitrary function f.

$$\sigma = f(V) \tag{4}$$

The optimal illumination is in the condition to maximize sharpness after adjusting the inputs of the light sources. Optimum algorithms are applicable to find the maximum sharpness within a short time. The optimum algorithms are derived to find the minimum of a cost function, so minus signs are given for the sharpness, which indicates the cost function. The following definition shows the expression in the optimum search of the cost function and the voltage inputs

$$\text{find} \quad \min \quad \rho = -\sigma = \text{g(V)} \quad \text{for} \quad \forall V \tag{5}$$

2.2 Random Search

The random search algorithm is composed of defining min-max vectors, obtaining Gaussian random, determining test points, updating current values and checking terminal conditions[38]. The concept of the random search and comparison with equal search is shown in figure 2. The minimum and maximum range of the voltage inputs is defined as follows for an N-sized vector, V.

$$\begin{aligned} V_{min} &= (V_{1min}, V_{2min}, \dots, V_{Nmin}) \\ V_{max} &= (V_{1max}, V_{2max}, \dots, V_{Nmax}) \end{aligned} \tag{6}$$

Gaussian random generators were proposed in some studies. Normal deviates can be written in the following equations[16]. The random value r in the PC is an unsigned integer, but uniform deviates between ± 1.0 are required in the generator, so r must be linearly scaled to $0 < |r'| < 1$. After this scaling operation, the sum of squares, r'_{sq}, can be calculated.

$$r'^2_{sq} = r'^2_1 + r'^2_2 \tag{7}$$

The Gaussian random r_g can be calculated through Box-Muller transformation as follows[34].

$$\begin{aligned} r'' &= \sqrt{-2ln(r'_{sq})/r'_{sq}} \\ r_g &= r'_1 * r'' \end{aligned} \tag{8}$$

The variation of the voltage inputs, Δ, is defined as follows.

$$\Delta = (\Delta_1, \Delta_2, \dots, \Delta_N) = (r_{g1}, r_{g2}, \dots, r_{gN}) \tag{9}$$

The variance was modified from that of the original random search for better convergence. As the iterations increase, the size of the variation necessary for convergence becomes smaller. So, we defined D to assist convergence, considering the current step k and convergence constant.

$$\begin{aligned} \Delta &= (r_{g1}, r_{g2}, \dots, r_{gN}) \times \frac{|\Delta_0|}{\eta |r_g|(k+1)} \\ &= (r_{g1}, r_{g2}, \dots, r_{gN}) \times D \end{aligned} \tag{10}$$

Boundary condition are applied to prevent excessive voltage input as follows.

$$\begin{aligned} V_i + \Delta_i > V_{max} &\rightarrow V_i + \Delta_i = V_{max} \\ V_i + \Delta_i < V_{min} &\rightarrow V_i + \Delta_i = V_{min} \end{aligned} \tag{11}$$

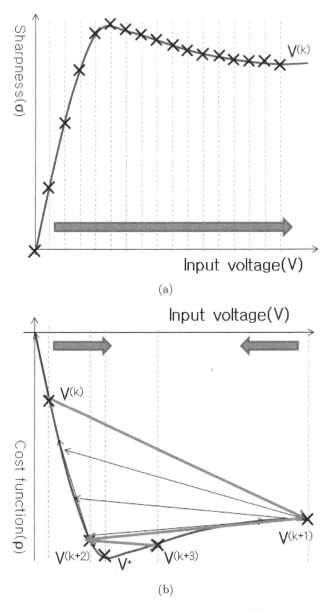

Fig. 2. Concept of random search for one variable example: (a) equal search (b) random search

The current inputs are updated when the current cost function is lower than that of the previous step.

$$V_{k+1} = V_k + \Delta_i \quad \text{if} \quad \rho(V_k + \Delta_i) < \rho(V_k) \tag{12}$$

After the new input is determined, the total number of iterations l can be updated. It is related to the convergence speed.

$$l = l + i$$
$$k = k + 1 \tag{13}$$

The terminal condition was checked for whether the current step is larger than a certain value, or whether the Δ is smaller than a minute value.

$$k > k_{max}$$
$$|\Delta_k| < \epsilon \tag{14}$$

3 Robust Parameter Design

3.1 Background

The proposed optimum algorithm can be used for multi-color auto-lighting and has parameters which can affect search performance. The algorithm parameters must be tuned to maximize search performance, and find the optimal illumination within a short time. The color mixing mechanism, the image processing and the random search are composed of nonlinear and complex functions, so it is difficult to build a tuning rule for the parameters. In actual cases of industrial machine vision, algorithm parameters are determined by trial-and-error, but many experiments are unavoidable to tune the parameters with questionable values. Robust design involves finding reliable parameters with minimal experiments. The Taguchi method is a famous type of experiment design and is widely used in industrial machine vision. Muruganantham applied the Taguchi method to calibrate a vision system considering illumination, ROI, and optical components[22]. The Taguchi method is applied to tune inspection algorithms, face recognition, and human gestures[27],[10],[5]. Illumination for lithography and LCD backlights are optimized by the Taguchi method[20,21]. Parameters in the Taguchi method are controllable factors which affect the performance of a target system. Larger sharpness is desirable for the image quality and fewer iterations are better for the searching time. Therefore, "larger the better" and "smaller the better" are applied for the sharpness and the iteration in the robust parameter design. The following are typical forms of "larger the better" and "smaller the better" in the Taguchi method.

$$SN = -10log\left(\frac{1}{n}\sum_{j=1}^{n}y_j^2\right) \tag{15}$$

$$SN = -10log\left(\frac{1}{n}\sum_{j=1}^{n}\frac{1}{y_j^2}\right) \tag{16}$$

In the equations, y_j is the performance index, such as sharpness and iteration, n is the number of experiments, and SN is the objective function for tuning.

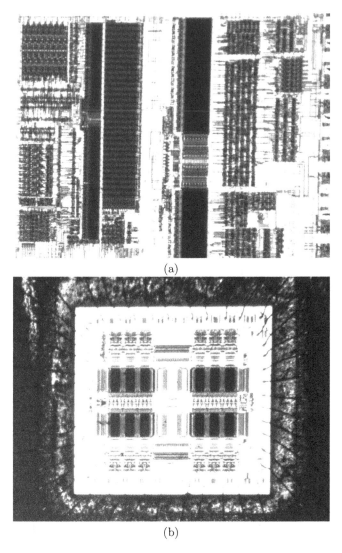

(a)

(b)

Fig. 3. Target patterns acquired by optimal illumination using simplex search: (a) pattern A (b) pattern B

3.2   Test Apparatus

Our test bed consisted of a vision camera, a coaxial lens, a manual stage, a color mixer, and a main PC. A framegrabber for digitizing images and a DAQ board for voltage inputs of the multiple sources were installed in the main PC. The voltage inputs were adjusted through DAC ports on the DAQ board. The internal connection of the color mixer for light transfer was based on optical fiber. Bundles of optical fibers from the multiple color sources were bound in a mixing chamber, and the mixed light was transferred to a coaxial lens[14].

RGB LEDs were used as the light sources. The mixed light illuminated a target pattern, and the camera acquired an image passing through the coaxial lens. The sharpness and the random search were computed on a PC. The voltage inputs of the multiple light sources were determined after the random search and sent to the DAC ports of the DAQ board. The LED AMPs generate driving current for the LEDs from the input current. The feedback loop was formed from the image acquisition to set the voltage inputs. The feedback loop was iterated by the random search until voltage inputs for the optimal illumination are found. The target patterns were commercial ICs (EP910JC35-ALTERA and Z86E3012KSE-ZILOG), as shown in figure 3.

3.3   Experiment Design for Random Search

The color mixer has three inputs for RGB colors, so the vector V is defined as an N=3 problem. For the control factors, the target parameters of Taguchi analysis were the RGB initial voltage (V_{R0}, V_{G0}, V_{B0}), the initial variance Δ_0, and the convergence constant η. The parameter tuning is aimed at decreasing the iterations and increasing the sharpness. The control factors and levels for the random search are summarized in table 1. A-E in the table are symbols of the control factors, and the number of levels is 5. So, the combination of the control factors and the levels is $5^5 = 3125$ cases, but the number in the Taguchi method is reduced to 25 cases. The orthogonal array for this problem is of $L_{25}(5^5)$, as shown in table 2. The analysis of the Taguchi method was focused on increasing the maximum sharpness and minimizing the number of iterations. The terminal condition was $k_{max} = 10$.

Table 1. Control factors and levels for random search

factors	code	Level				
		1	2	3	4	5
V_{R0}: Initial V_R	A	0.5	1.0	1.5	2.0	2.5
V_{G0}: Initial V_G	B	0.5	1.0	1.5	2.0	2.5
V_{B0}: Initial V_B	C	0.5	1.0	1.5	2.0	2.5
Δ_0: Initial variance	D	0.2	0.4	0.6	0.8	1.0
η: convergence constant	E	0.2	0.5	1.0	1.5	2.0

4   Results

4.1   Experimental Results

The optimal illumination found by equal search with $\Delta = 0.1$ was $\sigma_{max} = 392.76$ at $V = (0, 0, 1.2)$ for pattern A and $\sigma_{max} = 358.87$ at $V = (1.0, 0, 0)$ for pattern B. The result of this equal search was utilized as a reference solution. 25 cases of experiment were conducted by the parameter combination of the $L_{25}(5^5)$

Table 2. Orthogonal array of $L_{25}(5^5)$ for patterns A and B

Run #	Control factors					Pattern A					Pattern B				
	A	B	C	D	E	σ_{max}	l	V_R	V_G	V_B	σ_{max}	l	V_R	V_G	V_B
1	1	1	1	1	1	390.69	137	0	0	1.253	358.44	190	0.968	0	0
2	1	2	2	2	2	388.85	51	0.538	0.33	0.821	351.78	24	0.637	0.313	0.000
3	1	3	3	3	3	382.45	30	0.034	0.711	0.711	258.95	30	0.094	0.781	1.536
4	1	4	4	4	4	354.66	23	0.020	1.139	1.429	224.60	26	0.042	1.059	2.039
5	1	5	5	5	5	325.27	32	0.000	1.655	2.123	174.02	32	0.352	2.347	1.382
6	2	1	2	3	4	389.00	48	0.612	0.009	0.770	334.44	32	0.572	0.204	0.493
7	2	2	3	4	5	361.37	27	0.073	0.974	1.327	251.43	39	0.522	1.012	0.825
8	2	3	4	5	1	-	-	-	-	-	356.68	62	1.057	0.000	0.017
9	2	4	5	1	2	314.73	36	0.759	1.314	2.203	183.09	33	0.325	1.406	2.472
10	2	5	1	2	3	338.13	43	0.916	2.295	0.000	232.77	31	0.615	2.073	0.018
11	3	1	3	5	2	386.75	59	0.000	0.000	1.816	338.19	67	0.325	0.000	0.786
12	3	2	4	1	3	320.56	36	1.278	0.810	2.069	189.45	39	1.610	0.649	1.832
13	3	3	5	2	4	304.68	27	1.318	1.213	2.148	164.29	34	1.337	1.167	2.255
14	3	4	1	3	5	328.22	30	1.230	1.732	0.270	206.17	27	1.322	1.152	0.405
15	3	5	2	4	1	390.54	365	0.000	0.000	1.363	357.67	101	0.890	0.000	0.009
16	4	1	4	2	5	331.79	27	1.726	0.362	1.825	197.98	29	1.816	0.309	2.005
17	4	2	5	3	1	-	-	-	-	-	340.14	56	0.298	0.127	0.637
18	4	3	1	4	2	377.77	40	0.578	0.000	1.814	317.52	55	1.861	0.003	0.000
19	4	4	2	5	3	360.05	36	2.437	0.032	0.530	216.93	26	1.794	1.344	0.000
20	4	5	3	1	4	271.85	28	2.030	2.251	1.510	134.28	31	1.814	2.409	1.931
21	5	1	5	4	3	326.25	44	2.869	0.082	0.442	198.16	29	1.593	0.474	1.931
22	5	2	1	5	4	368.85	31	2.218	0.039	0.442	284.91	29	2.112	0.191	0.027
23	5	3	2	1	5	295.48	28	2.481	1.412	0.874	156.22	30	2.497	1.295	0.980
24	5	4	3	2	1	360.21	43	0.051	2.209	0.064	273.09	44	0.000	0.0030	3.221
25	5	5	4	3	2	296.74	39	3.010	1.849	0.000	158.70	30	0.750	1.753	2.385

orthogonal array. The values of σ_{max}, k, V_R, V_G, and V_B at the terminal state are shown in table 2. The σ_{max} values in the table were varied considerably, and the solutions were different from that of equal search. For the case of higher σ_{max}, the solutions approached the results of equal search but showed a larger number of iterations. So, the image fineness and searching time were counterparts when tuning the random search algorithm for optimal illumination. It failed to find optimum in cases #8 and #17.

4.2   Taguchi Analysis

As stated, the parameter design is aimed at increasing σ_{max} and decreasing l for quick optimal illumination. A "larger the better" approach was applied to σ_{max} and a "smaller the better" approach was applied to l. Taguchi analysis was carried out using MINITAB, a statistics package tool.

Figure 4 shows the result of Taguchi analysis and the influence of control factors for pattern A. Figure 4 (a) shows the effect of parameter on σ_{max}. The parameters of A, B, C, and E was decreased and that of D increased. So, a low initial

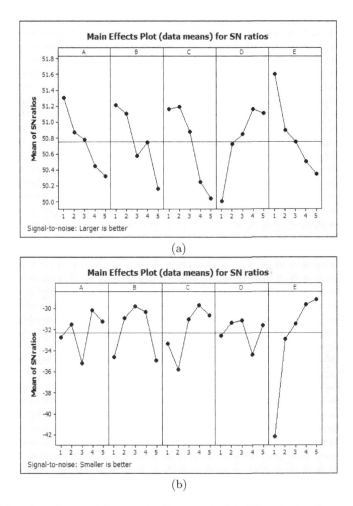

Fig. 4. SN ratios of control factors for Pattern A: (a) SN ratio for sharpness (b) SN ratio for total number of iterations

voltage and low η are recommended, but higher values are required for Δ_0. The most sensitive factor was η. Figure 4 (b) shows the effect of the parameter on l and represents the convergence speed. The parameters of C and E are increased but the others do not. The most sensitive parameters are E, and η. This indicates that σ_{max} and l should be balanced when the parameter E is selected. If the speed became too high, the solution was far from the equal search, so the parameter should be selected mainly by accuracy. A new combination of parameters for pattern A of $A_1B_2C_3D_5E_3$ was applied to a retest. The solution found by the retest was $(0,0,1.332)$ with $\sigma_{max} = 391.1$ and $l = 39$. Compared with $A_1B_1C_1D_1E_1$ and $A_3B_5C_2D_4E_1$, the sharpness was slightly smaller than that of the reference solution, but the optimal illumination was found with fewer iterations.

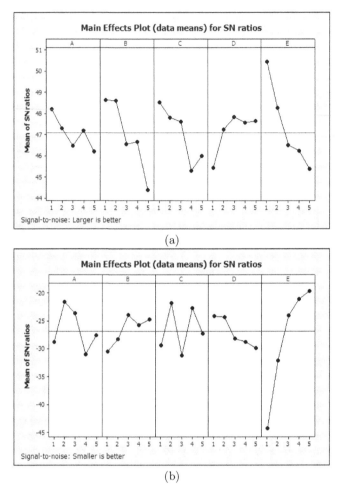

Fig. 5. SN ratios of control factors for Pattern B: (a) SN ratio for sharpness (b) SN ratio for total number of iterations

The trend of pattern B was similar to that of pattern A. Figures 5 (a) and (b) show the analysis results of σ_{max} and l, respectively. A, B, C, and E were decreased, and D was increased for σ_{max}. For l, B, and E were incremented, D was decremented, and the others showed no tendency. The most sensitive parameter was E, which should be balanced between the image quality and convergent speed. $A_2B_1C_1D_4E_3$ was selected for a new combination of parameters in the retest. The solution was $(0.624, 0.050, 0.229)$, which was different from the reference solution. However, σ_{max} was 357.45 and l was 23, which indicates that the image quality was almost the same as in the best case with fewer iterations.

Although the voltage inputs for optimal illumination after the Taguchi method were not equal to those of the reference solution, the difference in sharpness between the reference solution and the retest was small with fewer iterations, so the Taguchi method can provide useful tuning parameters with minimum experiments for the random search. The Taguchi method also supplies a basis for setting algorithm parameters, so parameter reliability is available in this optimal illumination problem.

5   Conclusion

Random search was introduced to the optimal illumination problem to acquire fine images in machine vision, and the Taguchi method was applied to tune the algorithm for the purpose of efficient use of the color mixer. The relation between the illumination and image sharpness was considered for N arbitrary sources. The Taguchi method was applied to tune the random search in consideration of higher sharpness and fewer iterations. The design parameters were the initial voltages, initial variation and convergence constant. 5 levels of 5 parameters were considered, so an $L_{25}(5^5)$ orthogonal array was constructed for Taguchi analysis. The parameter design was mainly focused on increasing sharpness and reducing iterations. The new parameter combination by the Taguchi method was retested and showed different results from the equal search. However, the retest results showed good image quality near the best cases with fewer iterations. The Taguchi method can provide reliable algorithm parameters for random search for quick optimal illumination with minimal experiments.

Acknowledgment. Acknowledgment This work was funded and supported by the Korea Institute of Industrial Technology. The authors are grateful to AM Technology (http://www. amtechnology.co.kr) for supplying RGB mixable color sources.

References

1. Arecchi, A.V., Messadi, T., Koshel, R.J.: Fieldguide to illimination, pp. 110–115. SPIE Press, Washington (2007)
2. Bueno-lbarra, M.A., Alvarez-Borrego, J., Acho, L., Chavez-Sanchez, M.C.: Fast autofocus algorithm for automated microscopes. Optic. Eng. 44(6), 063601 (2005)
3. Chern, N.K., Neow, P.A., Ang, M.H.: Practical issues in pixel-based autofocusing for machine vision. In: Proceeding of International Conference on Robotics and Automation, pp. 2791–2796 (2001)
4. Chi, C., Yoo, H.J., Ben-Ezra, M.: Multi-spectral imaging by optimized wide band illumination. Int. J. of Comp. Vis. 86, 140–151 (2010)
5. Chian, T.W., Kok, S.S., Lee, S.G., Wei, O.K.: Gesture based control of mobile robots. In: Proceeding of the Innovative Technogies in Intelligent System and Industrial Application, pp. 20–25 (2008)
6. Esparza, D., Moreno, I.: Color patterns in a tapered lightpipe with RGB LEDs. In: Proceeding of SPIE, vol. 7786 (2010)
7. Ding, F.J., Chen, Y.R., Chao, K.L.: Application of color mixing for safety and quality inspection of agricultural products. In: Proceeding of SPIE, vol. 5996, p. 59960R (2005)

8. Ding, F.J., Chen, Y.R., Chao, K.L., Kim, M.S.: Three-color mixing for classifying agricultural products for safety and quality. App. Opt. 45(15), 3516–3526 (2006)
9. Groen, F.C.A., Young, I.T., Ligthart, G.: A Comparison of different focus functions for use in autofocus algorithms. Cytometry 6(2), 81–91 (1985)
10. Ho, S.Y., Huang, H.L.: Facial modeling from an uncalibrated face image using flexible generic parameterized facial models. IEEE Trans. on Sys. Man and Cyber. - Part B 31(5), 706–719 (2001)
11. Kehtarnavaz, N., Oh, H.J.: Development and real-time implementation of a rule-based auto-focus algorithm. Real-Time Im. 9, 197–203 (2003)
12. Kim, H.T., Kang, S.B., Kang, H.S., Cho, Y.J., Kim, J.O.: Optical distance control for a multi focus image in camera phone module assembly. Int. J. Prec. Eng. and Manuf. 12(5), 805–811 (2011)
13. Kim, H.T., Kim, S.T., Cho, Y.J.: A review of light intensity control and quick optimum search in machine vision. In: Proceeding on International Symposium of Optomechatronic Technologies (2012)
14. Kim, H.T., Kim, S.T., Cho, Y.J.: An optical mixer and RGB control for fine images using grey scale distribution. J. of Optomech. 6(3), 213–225 (2012)
15. Kim, H.T., Kim, S.T., Kim, J.S.: Mixed-color illumination and quick optimum search for machine vision. J. of Optomech. 7(3), 208–222 (2013)
16. Knuth, D.E.: The art of computer programming, 2nd edn., vol. 2, pp. 116–122. Addison-Wesley, Massachusetts (1981)
17. Krotkov, E.P.: Active computer vision by cooperative focus and stereo. Springer, New York (1989)
18. Kuo, C.F.J., Chiu, C.S.: Improved auto-focus search algorithms for CMOS image-sensing module. J. of Infor. Sci. and Eng. 27, 1377–1393 (2011)
19. Lee, M.H., Seo, D.K., Seo, B.K., Park, J.I.: Optimal illumination stectrum for endoscope. In: Korea-Japan Joint Workshop on Frontiers of Computer Vision (2011)
20. Li, M., Milor, L., Yu, W.: Development of optimum annular illumination: A lithography-TCAD approach. In: Proceeding of Advanced Semiconductor Manufacturing Conference and Workshop, pp. 317–321 (1997)
21. Lin, C.F., Wu, C.C., Yang, P.H., Kuo, T.Y.: Application of Taguchi method in light-emitting diode backlight design for wide color gamut displays. J. of Disp. Tech. 5(8), 323–330 (2009)
22. Muruganantham, C., Jawahar, N., Ramamoorthy, B., Giridhar, D.: Optimal settings for vision camera calibration. Int. J. of Manuf. Tech. 42, 736–748 (2009)
23. Muschaweck, J.: Randomized micro lens arrays for color mixing. In: Proceeding of SPIE, vol. 7954 (2011)
24. Muthu, S., Gaines, J.: Red, Green and blue LED-based white light source: Implementation challenges and control design. In: Proceeding on the IEEE Industry Application Conference, vol. 1, pp. 515–522 (2003)
25. Proefschrift: Chemometrics in multispectral imaging for quality inspection of postharvest products. Thesis on Ph.D., aan de Radboud Universiteit Nijmegen (2005)
26. Santos, A., De Solorzano, C.O., Vaquero, J.J., Pena, J.M., Malpica, N., Pozo, F.D.: Evaluation of autofocus functions in molecular cytogenetic analysis. J. of Micros. 188, 264–272 (1997)
27. Su, T.L., Chen, H.W., Hong, G.B., Ma, C.M.: Automatic inspection system for defects classification of stretch kintted fabrics. In: Proceeding of Wavelet Analysis and Pattern Recognition, pp. 125–129 (2010)
28. Subbarao, M., Choi, T., Nikzad, A.: Focusing techniques, Tech. Report 92.09.04. State University of New York at Stony Brook (1992)

29. Sun, Y., Duthaler, S., Nelson, B.J.: Autofocusing in computer microscopy: Selecting the optimal focus algorithm. Micros. Res. and Tech. 65(3), 139–149 (2004)
30. Sun, C.C., Moreno, I., Lo, Y.C., Chiu, B.C., Chien, W.T.: Collimating lamp with well color mixing of red/green/blue LEDs. Opt. Exp. 20(S1), 75–84 (2011)
31. Tian, L., Slaughter, D.C., Norris, R.F.: Outdoor Field Machine Vision identification of tomato seedlings for automated weed control. Transactions of the ASAE 40(6), 1761–1768 (1997)
32. van Gorkom, R.P., van As, M.A., Verbeek, G.M., Hoelen, C.G.A., Alferink, R.G., Mutsaers, C.A., Cooijmans, H.: Etendue conserved color mixing. In: Proceeding of SPIE, vol. 6670 (2007)
33. Vriesenga, M., Healey, G., Sklansky, J.: Colored illumination for enhancing discriminability in machine vision. J. of Vis. Comm. and Im. Rep. 6(3), 244–255 (1995)
34. Press, W.H., Teukolsky, S.A., Vetterling, W.T., Flannery, B.P.: Numerical recipes in C, 2nd edn., pp. 289–290. Cambridge University Press, New York (1992)
35. He, X.F., Fang, F.: Flat-Panel Color Filter Inspection. Vis. Sys. Des., 20–22 (May 2011)
36. Yang, G., Nelson, B.J.: Micromanipulation contact transition control by selective focusing and microforce control. In: Proceeding of IEEE International Conference on Robotics and Automation, pp. 3200–3206 (2003)
37. Yap, P.T., Raveendran, P.: Image focus measure based on Chebyshev moments. Proceeding of Vision, Image and Signal Processing 151(2), 128–136 (2004)
38. Zabinsky, Z.B.: Random search algorithms. The Wiley Ency. of Oper. Res. and Manag. Sci. (2009)

Comparison of 3D Texture-Based Image Descriptors in Fluorescence Microscopy

Tomáš Majtner and David Svoboda

Centre for Biomedical Image Analysis,
Masaryk University, Brno, Czech Republic
{xmajtn,svoboda}@fi.muni.cz

Abstract. In recent years, research groups pay even more attention on 3D images, especially in the field of biomedical image processing. Adding another dimension enables to capture the entire object. On the other hand, handling 3D images also requires new algorithms, since not all of them can be modified for higher dimensions intuitively. In this article, we introduce a comparison of various implementations of 3D texture descriptors presented in the literature in recent years. We prepared an unified environment to test all of them under the same conditions. From the results of our tests we came to conclusion, that 3D variants of LBP in the combination with k-NN classifier are a very strong approach with the classification accuracy more than 99% on selected group of 3D biomedical images.

Keywords: 3D images, Texture descriptors, Fluorescence microscopy, Local Binary Patterns.

1 Introduction

Image descriptors still attract a lot of attention in many parts of image processing. They convert large amount of image data into a relatively short vector of numbers. Depending on the properties of the particular descriptor, these vectors capture different image characteristics. This is often used to distinguish between images and it can help us to decide, which images are visually similar.

Mapping of images to short vectors cannot be one to one as it naturally produces conflicts. Nevertheless, the properly defined image descriptors minimize the multiplicity of such conflicts. The ability of generating different vectors for different images is closely associated with the quality of selected image descriptors.

One can find various types of image descriptors. Traditional division is to the local and global ones. The former class is based on local features of images. They are searching for so called keypoints, i.e. interesting points in the image like corners or local minimum and maximum of intensity. Inspecting the neighborhood of these keypoints leads to generating of the feature vector. A representative of this approach is for example Scale-Invariant Feature Transform (SIFT) [18].

R.P. Barneva, V.E. Brimkov, and J. Šlapal (Eds.): IWCIA 2014, LNCS 8466, pp. 186–195, 2014.
© Springer International Publishing Switzerland 2014

Global image descriptors represent the other class, where entire image is used to extract feature vector. An example of this class is the group of MPEG-7 descriptors [20].

Based on the different image characteristics, further division can be employed in global descriptors to form approaches based on color, shape, texture, etc. In the field of fluorescence microscopy, we usually use the descriptors based on texture characteristics. It is because of their ability to capture inner structure of cells. This approach can be combined with the shape descriptors to enhance discrimination power. First, the boundary of the examined cell is determined by the shape descriptor and subsequently texture descriptor is used to describe inner part of the cell. The combination of these two approaches is often used implicitly because the background of the image degrades the power of texture descriptors.

Since the focus of research in fluorescence microscopy is still more concentrating on 3D images, in this paper, we offer a comparison of different texture-based descriptors on selected group of 3D fluorescence microscopy images. Our aim is to compare various approaches presented in the literature under the same conditions. Let us emphasize here, that by 3D image we understand full 3D volumetric representation of the image, not the 2D time series. In the article, we used the texture-based descriptors only and apply them on the images in the spatial domain, which is the main difference from the similar comparisons like [9].

This article is organized as follows: in the next section, we will provide brief summary of the texture-based approaches, that were employed in biomedical image analysis in recent years. The focus will be put on those already expanded for volumetric images. The third section will present the tested approaches with detailed discussion about setting up the parameters. The fourth section will be devoted to dataset description and classification. The achieved results of our tests will be presented in the end of the fourth section. The fifth section will discuss these results and emphasize the main observations. In the last section, we will conclude the main points and contribution of the article.

2   State of the Art

First notable publication dealing with texture-based image description was published in 1962 by Hu [15]. He introduced the theory of moment invariants for planar geometric figures and discussed both theoretical formulation and practical models of visual pattern recognition. The invariance under translation, similitude and orthogonal transformations were derived.

Some authors followed this approach (e.g. Kotoulas [17] and Flusser [10]) and new features based on this concept were derived. Zernike features [32] based on Zernike polynomials, which are orthogonal polynomials on unit disk, are an example of such descriptors. Their most important characteristics involves invariation to rotation and scale. It has been also shown, that the second and the third orders of Zernike features are equivalent to Hu's moments, when expressing them in terms of geometric moments [32]. The extension for 3D images was shown by Novotni and Klein [21].

Another approach is known as Haralick features. They were introduced in 1973 by Haralick et al. [14] and consist of 14 textural characteristics derived from so called co-occurrence matrix. Hence they are sometimes also referred as Gray-Level Co-occurrence Matrix (GLCM) statistics. This descriptor represents now the standard approach of the texture characterization and it was used by various biomedical research groups [3,29].

The extension of Haralick features for 3D volumetric images was presented in 2003 by Chen et al. [4]. The authors achieved 98% overall accuracy in the classification of HeLa cells with this approach. Further elaboration of 3D extension was presented in 2007 by Tesar et al. [33].

When continuing in the chronological order, next texture-based descriptor appearing in the literature was presented by Tamura et al. [30] in 1978. Here, the six features that correspond to human visual perception were introduced: coarseness, contrast, directionality, line-likeness, regularity and roughness. The main observation of the article was that coarseness, contrast and directionality are essential characteristics of texture and have high potential to be used as texture discriminators. Therefore, only this subset was further considered as so called Tamura features and subsequently incorporated into the various biomedical image recognition systems [24,27]. An extension of these features for 3D biomedical images was presented in 2012 [19].

A very popular approach named Local Binary Patterns (LBP) was introduced by Ojala et al. [23]. The original LBP operator works in 3×3 neighborhood, where central pixel value is used as a threshold for the other 8 pixels. The thresholded values are taken in clockwise order from top-left corner to create an 8-digit binary number, which represents the pattern. These patterns are usually converted to decimal and used to form a histogram that represents the feature vector. Currently, many different variants of LBP descriptor exists. For biomedical image analysis, most interesting ones are Median Binary Patterns [13] and Local Ternary Patterns [31] for their ability to handle noisy images.

The principal problem with extension of 2D LBP into 3D LBP is to find proper sampling of neighboring point in 3D. There has been presented version of LBP for spatiotemporal domain, which is a bit confusingly named Volume Local Binary Patterns (VLBP) [34]. The same authors presented also simplified version computing with three orthogonal planes (LBP-TOP) [35]. For 3D volume data, there has been two studies regarding the possibility of rotation invariance in 3D [8,1]. An interesting comparison between 2D and 3D LBP methods was presented in [25].

There can be found more texture-based methods for image characterization like Gabor filters [11] and Gauss-Markov model [26]. However, to our best knowledge, their usage for 3D biomedical image analysis was not examined yet.

In the next section, we will focus on the detailed description of 3D methods that we tested, with discussion of their parameters settings.

3   The Methods under the Scope

For the comparison, we decided to use 10 variants of 3D texture-based image descriptors.

Moments. The first of them is based on the theory of moments [10], which was referred in the beginning of the previous section. The descriptor consists of the 2nd, 3rd and 4th central moment, which are also known as the variance, skewness and kurtosis. We consider intensity values from the image voxels as the input set and calculate these well known statistics. The output is feature vector of length three.

Haralick Features. Another two descriptors are based on 3D Haralick features. In case of this descriptor, there are two significant parameters, that influence the result. First is the quantization level q. This parameter indicates, how many intensity values from original image will be merged together in co-occurrence matrix. In our experiments, the two highest classification accuracies were achieved for the quantization level equal to 4 and 8. The second important parameter of Haralick features is the distance between neighborhood values, which is used to form the co-occurrence matrix. In our case, the highest accuracies were achieved for this distance equal to 1. Therefore, we ended up with the two best configurations of Haralick features further referred as $Haralick_{q=4,d=1}$ and $Haralick_{q=8,d=1}$. Both had feature vector of length 14, based on the 14 characteristics derived from co-occurrence matrix.

Tamura Features. The next descriptor is a group of 3D Tamura features. This descriptor has no explicit parameters and consist of evaluation of three characteristics: coarseness, contrast and directionality. Each characteristic is represented by a single number, therefore the feature vector length in this case equals to three.

Local Binary Patterns. Last six tested 3D descriptors are all based on the idea of LBP and try to address the problem of sampling of neighboring points. Despite the fact, that VLBP and LBP-TOP were originally designed for spatiotemporal domain, we made a decision to try them in our tests. We did not need to modify them in any way, since the time domain can be considered as third dimension in volumetric images and handled the same way. The reason for our decision was based on the fact, that their design still holds the original idea of LBP and describe the 3D pattern around the central point in reasonable way also for volumetric data. VLBP takes three immediate frames to derive the pattern. There is a crucial common parameter P, which indicates number of sampling points in each 2D plane of the pattern. Large P leads to a long histogram, while short P means loosing more information. LBP-TOP is different in using orthogonal planes rather than the parallel ones where central pixel is included only in the middle frame. LBP-TOP also considers the feature distribution from each separate plane and concatenates them together, which leads to shorter vector when comparing with VLBP. In case of VLBP, we use the best performing variant $VLBP_{1,4,1}$ described in [34], where first parameter originally indicates the time interval between 2D slices. In our case of volumetric images it indicates, that we use immediate 2D slices. The second parameter is the number of local neighboring points around the central pixel and the third one is the radius. Since the feature vector length in this descriptor is derived from the expression 2^{3P+2} [34], in our case it is equal to 16384. In the LBP-TOP descriptor, we considered two parameters: the number of local neighboring points around the central pixel (P) and the radius (R). The best results was for $P = 8$ and $R = 1$. The feature vector of LBP-TOP is a concatenation of three histograms, each containing $2^8 = 256$ bins, which leads to the feature vector length of 768.

The remaining four descriptors based on LBP paradigm use the vertexes of platonic solids inscribed in the unit sphere to extract the patterns from the image. This idea is inspired by the article [1], where the authors used icosahedron to approximate equidistant sampling on the sphere. Full equidistant sampling on a sphere is a problematic task also known as Fejes Toths problem [7]. The authors of [1] used further splitting of icosahedron to better approximate the structure of the sphere. They ended up with large number of points, which leads to exponential increase in feature vector length, therefore they need to solve the merging of histogram bins. In our tests, we decide to examine four pattern samplings based on four platonic solids, namely tetrahedron, hexahedron, octahedron and icosahedron. The feature vector length for each of them is equal to 2^N, where N is number of vertexes of particular platonic solid. We do not consider dodecahedron, since it has 20 vertexes and will lead to the feature vector length of 2^{20}, which is approximately equal to the total number of voxels in images used for testing.

In the end, we added also one 2D descriptor, namely the original version LBP [22]. Reason for that was to examine, whether the inclusion of information from the third dimension really brings significant improvement in discriminating power. To be able to use 2D descriptor on 3D data, we apply the descriptor on each 2D slice of the 3D image and collect all the patterns from the same 3D image to common histogram. This idea leads to the feature vector length of 256, since LBP uses 2^8 histogram bins.

The following section will discuss tested dataset, describe the used classification procedure and introduce our results.

4   Data Preparation

We evaluate the presented methods on the database of the three different classes of biomedical images. First two, granulocyte and HL-60, represents the entire cell nuclei of particular type. They were all stained with DAPI and acquired with confocal microscope Zeiss 200M, Zeiss Plan-Apochromat 100x/1.40. to get full 3D volumetric representation. The third class, chromocenter, captures the inner parts of cell nuclei and was acquired by Leica TCS SP-5 X. Different types of the microscope ensure, that tested dataset does not suffer from the properties of only one particular acquisition system. The total number of images in the database is 269.

The original images were anisotropic, since during the process of image acquisition, the Z axis was acquired with different resolution than X and Y axis. This is a quite common case in biomedical image analysis. Therefore, prior to the feature extraction, we resampled all the original input images. We also unmasked the images to apply the descriptors only to the interesting parts and to avoid the background. In order to obtain the pre-segmented image data, we used Chan-Vese segmentation via graph cuts [5]. Fig. 1 shows an example of one database entry with corresponding mask derived by the segmentation method.

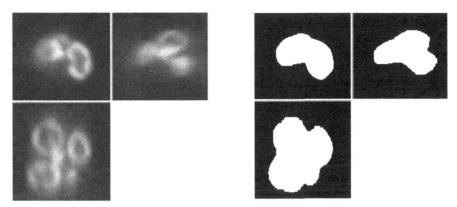

Fig. 1. The left part of the figure is granulocyte nuclei image from the database before unmasking. The right part of the figure is calculated mask of this image. Each parts consist of three individual images: the top-left image shows a selected xy-slice, the top-right image corresponds to yz-slice and the bottom one to xz-slice.

In this comparison study, we use k-Nearest Neighbor (k-NN) classifier implemented using MESSIF framework [2]. Please note, that for other classification methods the results could be different. k-NN was chosen for its highest accuracy in 2D experiments on similar images from fluorescence microscope [12].

For each descriptor in our classifier, features from all images are extracted and feature space with appropriate dimensionality is built. As a distance function we use the well-known L1 metric for all descriptors. The classification performance is computed using leave-one-out cross-validation method, where the entire database is considered.

To minimize the influence of chosen classification technique, we present the classification performance for different number of nearest neighbors, i.e. the different k's. The estimated class for each query image is chosen according to the weighted voting of the nearest neighbors, as it was proposed in [28].

5   Results and Discussion

There are several observations from the results presented in the Table 1, that we want to address here.

Table 1. The performance of the k-NN classification of selected 3D biomedical images based on different image descriptors and for different number of nearest neighbors (k). The tested dataset contains 269 images of 3 classes. In the table, FVL indicates the feature vector length.

Method	FVL	k=1 (%)	k=8 (%)	k=16 (%)
Moments	3	96.28	95.54	95.91
Haralick$_{q=4,d=1}$	14	92.57	94.80	93.31
Haralick$_{q=8,d=1}$	14	93.68	95.91	95.17
Tamura	3	89.59	92.19	92.19
VLBP	16384	98.88	**99.26**	**99.26**
LBP-TOP	768	**99.26**	98.88	97.77
3D LBP$_{tetrahedron}$	16	92.19	91.45	87.73
3D LBP$_{hexahedron}$	256	98.88	97.40	94.80
3D LBP$_{octahedron}$	64	**99.26**	98.51	97.77
3D LBP$_{icosahedron}$	4096	98.88	98.14	96.28
2D LBP	256	86.62	83.64	82.53

Firstly, all of the 3D descriptors exceed for at least two different examined nearest neighbors values the accuracy of 90%. In case of 2D LBP, we observe relatively worse results, which supports the claim, that the dependence among 2D slices is important and should be also considered. 2D descriptors applied to 3D image lack this feature and therefore are not sufficient.

Secondly, the achieved results are much higher than those presented in similar articles with concentration on 2D images only (e.g. [6]). Even from the other studies of 3D descriptors on biomedical images [8] we see, that the classification accuracy in 3D is very high. Hence, it is definitely better to work with 3D images when higher classification accuracy is needed.

Thirdly, the principle of LBP extended to 3D images by using various neighborhood on the sphere around the central pixel is very promising approach for biomedical image recognition. Different variants of LBP in 3D outperforms the other tested approaches. The only exception is 3D LBP$_{tetrahedron}$, where the neighborhood of central pixel is represented by only 4 points. This is clearly not sufficient for a proper description of a texture pattern in three dimensions and it have negative impact on the classification accuracy.

Fourthly, when increasing the value of k, the classification accuracy of LBP-based descriptors starts to decrease. This indicates, how the feature space, created from the feature vectors, is divided among examined classes. When we consider the nearest neighbor classifier ($k=1$), we require only the immediate neighbor to be of the same class as the query image for the correct classification. With the increasing value of k, more neighbors are considered and classifications with higher k tell us more about the clustering capacity of examined descriptor. Therefore, the descriptors with non-decreasing classification accuracy like Tamura features are also interesting, since they are less dependent on the chosen classifier parameters.

Fifthly, the results of VLBP are slightly surprising, since high dimensionality of feature vector is considered as problematic issue and dimensionality reduction is often employed [16]. Moreover, together with the LBP-TOP descriptor they represent the two best descriptors from our tests. Although they were introduced for 2D time series, their design of 3D pattern is applicable with high effectiveness also on tested volumetric biomedical images.

Sixthly, the Tamura features turn out to be less efficient than other tested 3D methods. Despite the fact, that their classification accuracy does not decrease with increasing number of nearest neighbors, the classification results are still below the other examined descriptors. One of the reasons we see in the fact, that Tamura features are non-parametric approach, which can not be tuned for specific requirements of examined dataset. On the other hand, descriptor based on central moments is also non-parametric and outperforms Haralick features for $k = 1$ and $k = 16$. This indicates us, that the distribution of intensity values in this kind of images is more significant than the co-occurrence of particular intensity values.

6   Conclusion

This paper deals with the topic of 3D image texture descriptors. It provides the review of the current state of the art approaches with the focus on the field of biomedical image analysis. The attention is concentrated on those texture-based descriptors, that were already extended for 3D images.

We presented a comparison of 10 individual descriptors based on 3D paradigm together with one 2D descriptor applied slice by slice on 3D images. All tested 3D descriptors achieved high accuracy level on tested dataset, when comparing them with results already presented on 2D biomedical datasets [6]. Especially the principle of LBP appears to be very efficient and should be considered as leading approach for 3D biomedical image recognition.

In the future, we would like to concentrate on the combining 3D descriptors inside the image classifier to achieve more robust system. This system will be used for more challenging datasets, that we expect in the following years.

Acknowledgement. This research was supported by the Czech Science Foundation (Projects P302/12/G157 and P202/14-22461S).

References

1. Banerjee, J., Moelker, A., Niessen, W.J., van Walsum, T.: 3D LBP-Based Rotationally Invariant Region Description. In: Park, J.-I., Kim, J. (eds.) ACCV 2012 Workshops, Part I. LNCS, vol. 7728, pp. 26–37. Springer, Heidelberg (2013)

2. Batko, M., Novak, D., Zezula, P.: MESSIF: Metric similarity search implementation framework. In: Thanos, C., Borri, F., Candela, L. (eds.) Digital Libraries: Research and Development. LNCS, vol. 4877, pp. 1–10. Springer, Heidelberg (2007)

3. Boland, M.V., Murphy, R.F.: A neural network classifier capable of recognizing the patterns of all major subcellular structures in fluorescence microscope images of hela cells. Bioinformatics 17(12), 1213–1223 (2001)

4. Chen, X., Velliste, M., Weinstein, S., Jarvik, J.W.: Location proteomics: building subcellular location trees from high-resolution 3D fluorescence microscope images of randomly tagged proteins. In: Storage and Retrieval for Image and Video Databases, vol. 4962, pp. 298–306 (2003)

5. Daněk, O., Matula, P., Maška, M., Kozubek, M.: Smooth Chan-Vese Segmentation via Graph Cuts. Pattern Recogn. Lett. 33(10), 1405–1410 (2012)

6. Doshi, N.P., Schaefer, G.: A comprehensive benchmark of local binary pattern algorithms for texture retrieval. In: 2012 21st International Conference on Pattern Recognition (ICPR), pp. 2760–2763. IEEE (2012)

7. Erdos, P., Pach, J.: On a problem of on a problem of L. Fejes Tóth. Discrete Mathematics 30(2), 103–109 (1980)

8. Fehr, J., Burkhardt, H.: 3D rotation invariant local binary patterns. In: 19th International Conference on Pattern Recognition, ICPR 2008, pp. 1–4 (2008)

9. Fehr, J.: Local Invariant Features for 3D Image Analysis: Dissertation. Suedwestdeutscher Verlag fuer Hochschulschriften, Germany (2009)

10. Flusser, J., Kautsky, J., Šroubek, F.: Implicit moment invariants. Int. J. Comput. Vision 86(1), 72–86 (2010)

11. Fogel, I., Sagi, D.: Gabor filters as texture discriminator. Biological Cybernetics 61(2), 103–113 (1989)

12. Foggia, P., Percannella, G., Soda, P., Vento, M.: Early experiences in mitotic cells recognition on hep-2 slides. In: IEEE 23rd International Symposium on Computer-Based Medical Systems (CBMS), pp. 38–43 (2010)

13. Hafiane, A., Seetharaman, G., Zavidovique, B.: Median Binary Pattern for Textures Classification. In: Kamel, M., Campilho, A. (eds.) ICIAR 2007. LNCS, vol. 4633, pp. 387–398. Springer, Heidelberg (2007)

14. Haralick, R.M., Shanmugam, K., Dinstein, I.: Textural features for image classification. IEEE Trans. on Systems, Man and Cyber. SMC-3(6), 610–621 (1973)

15. Hu, M.K.: Visual Pattern Recognition by Moment Invariants. IRE Transactions on Information Theory IT-8, 179–187 (1962)

16. Jaganathan, Y., Vennila, I.: Feature dimension reduction for efficient medical image retrieval system using unified framework. J. Comput. Sci. 9, 1472–1486 (2013)

17. Kotoulas, L., Andreadis, I.: Image Analysis Using Moments. In: 5th Int. Conf. on Technology and Automation, pp. 360–364 (2005)

18. Lowe, D.: Object recognition from local scale-invariant features. In: The Proc. of the 7th IEEE Int. Conf. on Computer Vision, vol. 2, pp. 1150–1157 (1999)

19. Majtner, T., Svoboda, D.: Extension of Tamura Texture Features for 3D Fluorescence Microscopy. In: Second Intern. Conf. on 3D Imaging, Modeling, Processing, Visualization and Transmission (3DIMPVT), pp. 301–307 (2012)

20. Manjunath, B., Salembier, P., Sikora, T. (eds.): Introduction to MPEG-7: Multimedia Content Description Interface. Wiley & Sons, Inc., New York (2002)

21. Novotni, M., Klein, R.: Shape retrieval using 3D Zernike descriptors. Computer Aided Design 36, 1047–1062 (2004)

22. Ojala, T., Pietikainen, M., Harwood, D.: Performance evaluation of texture measures with classification based on kullback discrimination of distributions. In: Proc.

of the 12th IAPR Intern. Conf. on Patt. Recog. - Conf. A: Computer Vision & Image Processing, vol. 1, pp. 582–585 (1994)

23. Ojala, T., Pietikainen, M., Maenpaa, T.: Multiresolution gray-scale and rotation invariant texture classification with local binary patterns. IEEE Trans. on Pattern Analysis and Machine Intelligence 24(7), 971–987 (2002)

24. Orlov, N., Eckely, D.M., Shamir, L., Goldberg, I.G.: Machine vision for classifying biological and biomedical images. In: Visualization, Imaging, and Image Processing (VIIP 2008), pp. 192–196 (2008)

25. Paulhac, L., Makris, P., Ramel, J.-Y.: Comparison between 2D and 3D Local Binary Pattern Methods for Characterisation of Three-Dimensional Textures. In: Campilho, A., Kamel, M. (eds.) ICIAR 2008. LNCS, vol. 5112, pp. 670–679. Springer, Heidelberg (2008)

26. Rellier, G., Descombes, X., Falzon, F., Zerubia, J.: Texture feature analysis using a Gauss-Markov model in hyperspectral image classification. IEEE Transactions on Geoscience and Remote Sensing 42(7), 1543–1551 (2004)

27. Shamir, L., Orlov, N., Eckley, D.M., Macura, T., Johnston, J.: Wndchrm – an open source utility for biological image analysis (2008)

28. Stoklasa, R., Majtner, T., Svoboda, D.: Efficient k-NN based HEp-2 cells classifier. Pattern Recognition (2013) (in press)

29. Svoboda, D., Kozubek, M., Stejskal, S.: Generation of digital phantoms of cell nuclei and simulation of image formation in 3D image cytometry. Cytometry A 75(6), 494–509 (2009)

30. Tamura, H., Mori, S., Yamawaki, T.: Textural features corresponding to visual perception. IEEE Tran. on Systems, Man and Cyber. 8(6), 460–473 (1978)

31. Tan, X., Triggs, B.: Enhanced Local Texture Feature Sets for Face Recognition Under Difficult Lighting Conditions. IEEE Transactions on Image Processing 19(6), 1635–1650 (2010)

32. Teague, M.R.: Image analysis via the general theory of moments. Journal of the Optical Society of America (1917-1983) 70, 920–930 (1980)

33. Tesar, L., Smutek, D., Shimizu, A., Kobatake, H.: 3D extension of Haralick texture features for medical image analysis. In: Proceedings of the Fourth IASTED International Conference on Signal Processing, Pattern Recognition, and Applications, SPPRA 2007, Anaheim, CA, USA, pp. 350–355. ACTA Press (2007)

34. Zhao, G., Pietikäinen, M.: Dynamic Texture Recognition Using Volume Local Binary Patterns. In: Vidal, R., Heyden, A., Ma, Y. (eds.) WDV 2005/2006. LNCS, vol. 4358, pp. 165–177. Springer, Heidelberg (2006)

35. Zhao, G., Pietikainen, M.: Dynamic Texture Recognition Using Local Binary Patterns with an Application to Facial Expressions. IEEE Trans. Pattern Anal. Mach. Intell. 29(6), 915–928 (2007)

Human Body Model Movement Support: Automatic Muscle Control Curves Computation

Jana Hájková[1] and Josef Kohout[1,2]

[1] Department of Computer Science and Engineering, University of West Bohemia,
Plzeň, Czech Republic
[2] New Technologies for Information Society (NTIS), University of West Bohemia,
Plzeň, Czech Republic
{hajkova,besoft}@kiv.zcu.cz

Abstract. In this paper we present a novel approach of an automatic computation of muscle control curves. It is based on skeletonization of a triangular surface mesh representing the muscle. Automatically determined control curves are then connected to the skeleton of the human body model so as to govern the deformation of the muscle surface when the skeleton moves. The method, which was implemented in C++ using VTK framework, was integrated into the human body framework being developed at our institution and tested on the walking lower limbs. The results show that the control curves produced by the method have a positive effect on the deformation and, therefore, are preferred to manually defined lines of action that are used as control curves in the human body framework at present.

Keywords: Muscle modelling, Musculoskeletal model, Line of action, Control curve, Skeletonization.

1 Introduction

Musculoskeletal modelling and simulation enables better prognosis and more accurate treatment for the patients suffering from various musculoskeletal disorders, e.g., from osteoporosis. As creating a patient-specific musculoskeletal model directly from medical images of each patient is normally not feasible due to the cost and human effort involved, a more efficient solution lies in constructing an atlas - a generic musculoskeletal model that is then deformed to fit a particular patient using specific individual features captured from data related to the target subject.

Atlas models in common clinical practice use represent a muscle by one or more polylines, named lines of action, joining the origin and insertion attachment areas of the muscle, i.e., the sites at which the muscle is attached to the bone by a tendon, along which the muscle mechanical action occurs. We note that usually no more than two lines of action per a muscle are specified as their specification requires user intervention.

An advantage of these approaches, which makes them so popular, is their rapid processing speed. However, representing a muscle by a set of lines of action

R.P. Barneva, V.E. Brimkov, and J. Šlapal (Eds.): IWCIA 2014, LNCS 8466, pp. 196–211, 2014.

provides very limited insight. This is probably why recently there has been call for visualization of physiologically correct positions and shapes of the muscles during motion. A straightforward approach is to extend existing musculoskeletal models as follows. A muscle is represented not only by a set of lines of action but also by a surface mesh that deforms in reaction to a change of lines of action, i.e., lines of action serves as control curves for the deformation; these are considered to be the muscle skeleton in deformation methods.

This paper investigates the usability of lines of action to be control curves and proposes a simple automatic method for computation of control curves that describe the muscle shape better. It is structured as follows. Line of action models are briefly described in related work in Section 2. Section 3 is dedicated to our musculoskeletal model and basics of muscle deformation method we use. Analysis of using lines of action as control curves is presented in Section 4. This section also describes the proposed method for an automatic computation of better behaved control curves and their usage. Results are discussed in Section 5. Section 6 concludes the paper.

2   Related Work

Our approach comes out from models, which are used in biomechanical practice and that represent muscles by a set of lines of action. When the muscle path is almost linear throughout a motion, the line of action can be defined as a simple straight line. Otherwise, it is considered to be a polyline passing through a number of intermediate via points, fixed to the underlying bone to move automatically with it. Different methods show how to describe the muscle path in a more realistic way [1], [7], [8] or to wrap around wrapping surfaces, again fixed to the underlying bone to geometrically constrain the line of action [2], [3], [5], [11], [12]. In the most sophisticated models of this kind, lines of action are considered as elastic strings that can wrap automatically around multiple surfaces, known as obstacles, which might be a set of spheres, cylinders, ellipsoids or cones [5].

3   Our Model

3.1   Model Atlas

Our data model, which is described in detail in [10], contains bones (Fig. 1a) and muscles (Fig. 1b), and description of their hierarchy and connections. Muscles and bones are represented by triangular surface meshes; each of them consisting of thousands of triangles. Each muscle is furthermore associated with one or more lines of action. Muscles that have large (e.g., gluteal muscles) or multiple (e.g., biceps femoris) attachment areas are associated with two lines of action, all other muscles with just one (Fig. 1c). Attachment areas are given as landmarks specified by an expert and fixed to the surface of the relevant bone to move automatically with it. An example of a muscle with its attachment areas and lines of action can be seen in Fig 2.

The data atlas is in fact a static model considered to be in the rest pose (RP) and only after it is fused with motion data, simulation can be provided. For each simulation time frame, called current pose (CP), the position and rotation of every bone, is available. As lines of actions are defined to be relative to bones, their paths automatically update.

a)   b)   c)

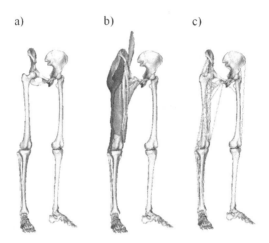

Fig. 1. Our musculoskeletal model in the rest pose (RP); a) bones; b) muscles; c) lines of action

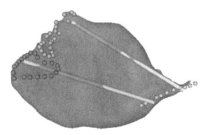

Fig. 2. Gluteus maximus with its attachment areas connecting muscle with bones (denoted as coloured spheres) and lines of action (yellow lines)

To get any muscle from RP into CP, the deformation of its surface mesh has to take its place. How the muscle surface should change to fit the anatomy at the current time frame is mostly determined by something we intuitively call "the main direction of the deformation", which can be defined as the difference between the control curves of the muscle in RP and their counterparts in CP. In the description that follows, lines of action serve as control curves, i.e., the main direction of the deformation is given by the difference between the new and previous paths of the line of action, unless explicitly said otherwise.

3.2 Deformation Method

Basic version of the deformation method we use is described in [9]. Three constraints are defined - preservation of the muscle shape, its main deformation direction and its volume. From these constraints the overconstrained linear system with non-linear boundary constraint of volume was mathematically derived. We solve the equations iteratively using Gauss-Newton method with Lagrange coefficients. Because muscles and bones lie close to each other, after the deformation the neighbouring meshes can be intersected. This is solved by local corrections provided after each deformation step. An example of gluteus maximus deformation is in Fig. 3.

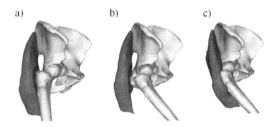

Fig. 3. Position of bones (pelvis and femur) and shape of the gluteus maximus: a) muscle and bones in RP; b) bones in CP, muscle in RP; c) muscle deformed to CP

To solve situation where bones are placed in CP and RP in large distance, the original method was extended and its interpolation version was described, see details in [6]. The input data and the output usage of the model remain the same.

4 Lines of Action and Control Curves

Our experiments show that the muscle deformation algorithm works correctly for all muscles with well-defined lines of action. In such cases, the line of action represents correctly the whole length of the muscle and its basic shape. An example of well-defined lines of action can be seen in Fig. 4.

Unfortunately, lines of action for many muscles are defined in such a way that they do not approximate the muscle shape enough. Very often the main problem is that the line of action controls only a part of the muscle and the rest remains uncontrolled, though from mechanical point of view it is defined correctly, or it is controlled by the existing line of action placed and shaped absolutely far off

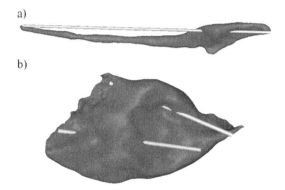

Fig. 4. Well defined lines of action: a) tensor fascia latae; b) gluteus maximus

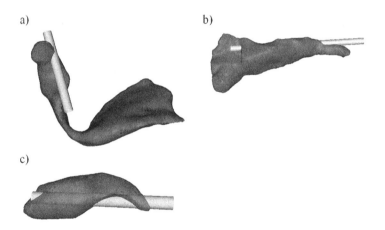

Fig. 5. Examples of badly defined lines of action: a) obturator internus; b) piriformis; c) quadratus femoris

(as for example in Fig. 5a). In other cases the line of action is too short or long (Fig. 5b) or it simplifies the shape of the muscle a lot (Fig. 5c).

Fig. 6 brings an example of the most complex problematic lines of action. The original line of action of illiacus in RP is represented by green poly-line, blue poly-line indicates the shape of line of action in CP. As can be seen, both lines of action can differ in shapes and also in count of point. Also this line of action does not represent the muscle shape (especially its top part) correctly.

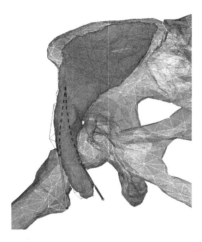

Fig. 6. The original line of action of illiacus in RP (green curve) and in CP (blue curve)

All lines of action described above are usable for biomechanical purposes. But if we want to extend the model for the visualization features (to enable to analyze the muscle behaviour in dependence on the model motion), we get into a problem when using lines of action to define the main direction of deformation. The muscles with inappropriate lines of action are not deformed correctly (see results). That is why we have decided to automatically compute a special curve for every single muscle that will better correspond to the muscle surface shape and will be applicable for the visualization of muscle deformation. The proposed method for computing such a control curve is described in the next section.

4.1   Control Curve Computation

We decided to use method of muscle skeletonization, which was in our system already prepared. Its complete detail and related work will be described in a separate paper that is being prepared. To understand the main idea, the basic approach will be shown now. For the sake of simplicity, let us assume that we process a muscle that is connected to two bones only, i.e., that each end of the muscle contains just one attachment area. When the bones are moving, the muscle has to keep the connection and to change its shape according to its main direction of deformation defined by the difference between its control curve in RP and CP poses. The control curve in RP is computed by the following process.

First, the number n of points P_i ($i \in [0, n]$) of the control curve has to be determined. It should be chosen according to the formula:

$$n = 2_i + 2, i \in N_0 \qquad (1)$$

where i represents the number of binary subdivisions of the muscle.

Providing that the boundaries of both attachment areas are present in the triangular surface mesh of the muscle (this can be easily ensured by projecting the expertspecified landmarks onto the surface of the muscle and inserting them into the triangulation), the vertices of the muscle surface mesh can be clustered according to their topological distance from boundaries both attachment areas into two clusters: one contains the part of mesh closer to the origin area, the other the part closer to the insertion.

Each cluster can be further divided in a similar way; with each step we get twice as many clusters as before (Fig. 7a). The partioning procedure ends when we get (n-2) clusters. We note that this approach is contingent on the assumption that the muscle surface is represented by the regular triangular mesh. In other cases, clustering must be based on geodetic distances, for which the harmonic scalar function [4] could be used.

For each cluster, its centroid is determined as a centre of gravity of all muscle mesh surface vertices belonging into the cluster. As can be seen in Fig. 7b, at this moment points P_1 to P_{n-1} are computed. Because the control curve should represent the muscle in its whole length, centroids are not sufficient vertices of the final poly-line representation. In dependence on the length of the muscle and number of computed clusters, there remain outer parts of the muscle without any control. That is why we have to extend the poly-line representation for two ending points (P_0 and P_n) lying at the ends of the muscle and to add them to the beginning and to the end of the precomputed poly-line.

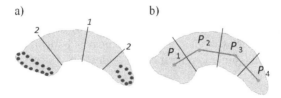

Fig. 7. Clusters computation: a) muscle and its attachment areas, computed clusters; b) clusters with their centroids forming the central part of the control curve

Each ending point should lie near the muscle surface and should be placed close to one of the furthest muscle ends. Of course, we could search this point as the furthest vertex of the muscle surface mesh (containing hundreds of vertices to be tested). To decrease the computational time, we decided to select one point from the attachment area representation (composed from units or tens of points). The computation is outlined in Fig. 8 on the example of P_0 detection. Among the attachment area points the furthest one to the outer cluster centroid is used. By this approach we would select the direction marked with the dotted line. As can be seen well, the direction of the final segment would not absolutely correspond with the shape of central part. To preserve the shape of the control curve, only such points are tested, which lie in the direction of the precomputed control curve ending segment and its surrounding (highlighted by the shadow triangle), that means that the line segments P_0P_1 and P_1P_2 make an angle of almost 180°. Finally, the control curve is composed from both ending points (P_0 and P_n) and all centroids computed in the previous step (P_1, ... P_{n-1}). The result can be seen in Fig. 9.

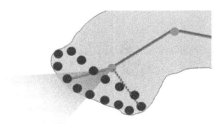

Fig. 8. Detail of selection of P_0 among attachment area points. Red dotted line represents selection of point without considering last segment direction.

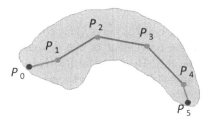

Fig. 9. Final control curve

4.2   Control Curve Motion

Each control curve is computed for the RP of the model. During the model motion, bones are changing their position and also they can rotate. Especially in the case of rotation we need to keep it also in the muscle deformation process and so a control point is computed for each control curve. It can be determined e.g. as the centre of vector connecting end points of the control curve.

Moreover, for each muscle connected to the moving bones also its control curve has to move to give the muscle the right direction of deformation all the time. That is why the control curve is fixed to the attached bones. First, for each end point, the nearest vertex of the bone surface mesh is chosen (Fig. 10a), and then the distance of nearest point and the appropriate control curve end point is computed (Fig. 10b). Index of the vertex and distance between both points are saved directly into the control curve structure. In this step the control curve is fixed to the bones, the original shape and number of points remain the same. The final motion of the control curve can be seen in Fig. 10c.

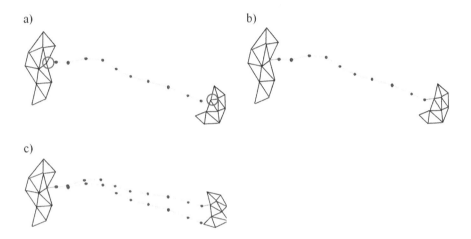

Fig. 10. Moving of the control curve: a) nearest bone surface vertices determination; b) end points fixing; c) new control curve position computation

When the new position of control curve, i.e., control curve in CP, is to be computed, the vectors of the movement of the nearest bone points are determined; let us call them (s_0, s_n). Because the RP distance has to be preserved, both control curve end points should move to their appropriate position, i.e., P_0 is moved in the direction of s_0 and analogically P_n moves in the direction of s_n. The other points $(P_1, ..., P_{n-1})$ are linearly interpolated to keep their relative distance to these end points and to preserve the original shape of the control curve. The position of all points can be computed by the equation:

$$\boldsymbol{P}'_i = \boldsymbol{P}_i + (\frac{n-i}{n-1}.\boldsymbol{s}_0 + \frac{i}{n-1}.\boldsymbol{s}_n), i \in < 0, n > \qquad (2)$$

An example of the movement of every single point is highlighted in detail in Fig. 11. The original control curve consists of points P_i, green dashed vertices represent motion of P_0, while blue dotted arrows symbolize P_3 movement. Two red arrows show the final direction for the other points. New (moved) control curve consists of points P'_i.

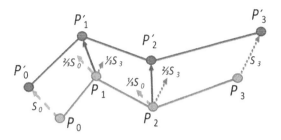

Fig. 11. Movement of control curve points

This method works best for curves which are relatively straight. For instance, if P_n moves into a large distance from P_0 and, moreover, P_0 does not move at all (see Fig. 12), in the final state of curve the distances between points does not change in the same way. For the further usage of control curve, however, this fact does not bring any problem.

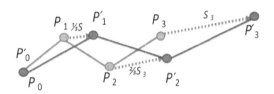

Fig. 12. Final shape of control curve after movement of one end point for a large distance

Other complications may arise for extremely curved curves (such as for obturator internus, Fig. 16). If the motion of one point is provided in a direction going from P_n to P_{n-1} or close and the distance is large, the shape of the curve is deformed (see Fig. 13). Fortunately, according to our best knowledge, curved muscles are placed in the human body in such positions that their end points do not allow to move for a long distances.

According to facts described above, this approach is for our experiments sufficient. Still, if problems appear, the spring system used in methods with elastic lines of action (described in Section 2) could be used.

5   Results

The method described above computes control curves composed of the predefined number of points. With respect to the approach, the final number of points is set as $2i + 2$ (where $i \in$ N, $i \neq 0$). The more points we use for the control curve, the more accurate approximation of muscle shape we get. As example, sartorius was selected because of its length and flexion in its bottom part. In the first case

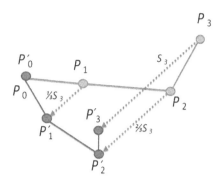

Fig. 13. Final shape of extremely curved control

(Fig. 14a), the segmentation of the muscle on clusters were provided five-times and so 32 centroids were computed. The whole control curve composes of 34 points. In Fig. 14b, the original mesh was divided only to four clusters and so the final poly-line consists of 6 points. It can be seen well that in the second case the bottom part is not approximated well and the control curve passes outside the muscle.

For the muscle deformation method we need to use input control curve with optimal number of segments. Each additional segment brings a new constraint into the over-constrained linear system used in the muscle deformation method mentioned in Section 3.2.

Because muscles in human body have different size, for each muscle the number of points should be determined individually. That is why we have implemented also automatic computation of the number of points. It is estimated in dependence on the length of the muscle, i.e., on the distance between furthest points of both attachment areas. This distance is divided into segments with length approximately 3-4 times larger than the surface triangle average side (that is 2-4mm in dependence on a particular muscle). Of course, in the case of curved shape of muscle, the distance is shorter than the real length of the muscle, but with regard to the square root number adjustment of control curve points, this fact does not cause fundamental problem. Fig. 15 shows dependence of time needed for the muscle deformation on the medial distance of control line points. Test was provided on two muscles of different size (iliacus and sartorius) and the result show that the deformation algorithm was most quick, when the distance od points was around 4,5mm. The exact values of measured time are not important, the trend of its progression should be explored.

In Fig. 5, several muscles with problematically defined lines of action in the data model were mentioned. Results of control curves computation for the same muscles (obturator internus, piriformis and quadratus femoris) can be found in Fig. 16. The curves go through the whole length of muscles and, if the proper number of points is used for the poly-line, the shape and curvature are approximated well.

a)   b)

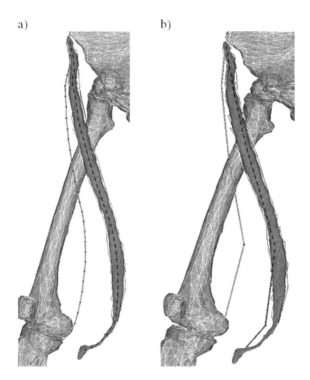

Fig. 14. Control curves computed for sartorius, poly-line consists of a) 34 points; b) 6 points

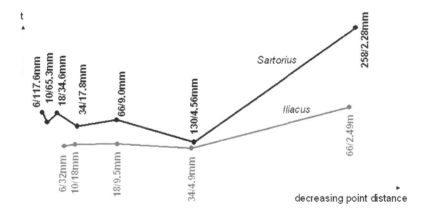

Fig. 15. Dependence of time needed for the muscle deformation on the medial distance of control curve points computed for iliacus and sartorius. Results are described with two values - number of coltrol curve points/medial distance of control curve points. Distance between points is decreasing while going to the right of the plot.

a)   b)

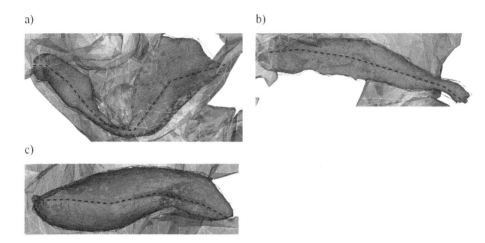

c)

Fig. 16. Results of automatic control curve computation for real muscles: a) obturator internus; b) piriformis; c) quadratus femoris

In Fig. 6 complicated original shape of the illiacus line of action was described. The same situation from the similar view can be found in Fig. 17. The control curve represents the muscle shape more sufficiently. The curving in the bottom part is not optimal, which is caused by the position of attachment area between illiacus and femur.

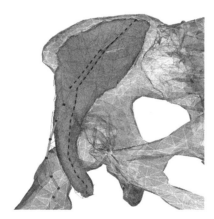

Fig. 17. Computed control curve of illiacus in RP (green curve with blue points) and in CP (blue curve with green points)

Following figures show result of illiacus deformation provided according to the original line of action (a) and to the automatic computed control curve (b) from the front (Fig. 18) and top (Fig. 19) view. Results can be compared with the basic muscle shape in RP in Fig. 17. During the deformation when lines of action are used, the top part is getting thinner and narrower and so it loses its volume. Because the deformation method preserves volume of the muscle, the bottom part prolongs incorrectly. Moreover, from the top view there can be seen the space between illiacus and pelvis that has arisen during the deformation. Results of deformation provided according to the control curve solve all these problems, despite the abovementioned one suboptimal part of the computed control curve.

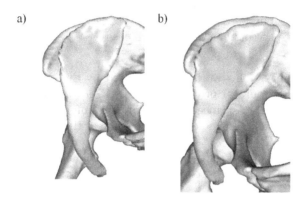

Fig. 18. Front view on illiacus deformed according to a) original line of action; b) computed control curve

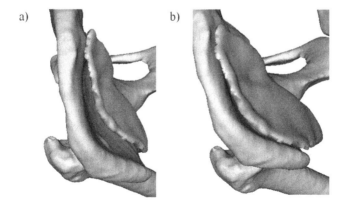

Fig. 19. Top view on illiacus deformed according to a) original line of action; b) computed control curve

6 Conclusions

We have developed method for an automatic computation of control curves. Our tests give promising visual results; the muscle deformation method works with our results well and all tested muscles were deformed correctly in contradistinction to deformation with original lines of action present in the data model. By using the automatic control curve computation, the human model does not need to rely on measured data, that are very often inaccurate and it is also possible to use model simulation in unmeasured human motion positions. In the theoretical part of this paper also muscles with more attachment areas which provide connection of the muscle to more than two bones were mentioned. Our future plan is to enhance the method also for such cases.

Acknowledgments. This work was supported by the NEXLIZ (New Excellence in Human Resources) project, Reg. No. CZ.1.07/2.3.00/30.0038, by the project SGS-2013-029 - Advanced Computing and Information Systems and by the Information Society Technologies Programme of the European Commission under the project VPHOP (FP7-ICT-223865). The authors would like to thank the people who contributed to the realisation of our human body framework and to various people who provided condition under which the work could be done.

References

1. Arnold, A.S., Salinas, S., Asakawa, D.J., Delp, S.L.: Accuracy of muscle moment arms estimated from MRI-based musculoskeletal models of the lower extremity. Computer Aided Surgery 5, 108–119 (2000)
2. Audenaert, A., Audenaert, E.: Global optimization method for combined spherical-cylindrical wrapping in musculoskeletal upper limb modelling. Computer Methods and Programs in Biomedicine 921, 8–19 (2008)
3. Delp, S.L., Loan, J.P.: A computational framework for simulation and analysis of human and animal movement. IEEE Computing in Science and Engineering 2(5), 46–55 (2000)
4. Dong, S., Kircher, S., Garland, M.: Harmonic functions for quadrilateral remeshing of arbitrary manifolds. Comput. Aided Geom. Des. 22(5), 392–423 (2005)
5. Garner, B.A., Pandy, M.G.: The obstacle-set method for representing muscle paths in musculoskeletal models. Computer Methods in Biomechanics and Biomedical Engineering 3(1), 1–30 (2000)
6. Hájková, J., Kohout, J.: Musculoskeletal system modelling – interpolation method for muscle deformation. In: Proceedings of International Conference on Computer Graphics Theory and Applications (GRAPP 2013), Spain (2013)
7. van der Helm, F.C., Veenbaas, R.: Modelling the mechanical effect of muscles with large attachment sites: application to the shoulder mechanism. Journal of Biomechanics 24(12), 1151–1163 (1991)
8. Jensen, R.H., Davy, D.T.: An investigation of muscle lines of action about the hip: A centroid line approach vs. the straight line approach. Journal of Biomechanics 8(2), 103–110 (1975)

9. Kohout, J., Clapworthy, G.J., Martelli, S., Viceconti, M.: Fast realistic modelling of muscle fibres. In: Csurka, G., Kraus, M., Laramee, R.S., Richard, P., Braz, J. (eds.) VISIGRAPP 2012. CCIS, vol. 359, pp. 33–47. Springer, Heidelberg (2013)
10. Kohout, J., Clapworthy, G.J., Zhao, Y., Tao, Y., Gonzales-Garcia, G., Dong, F., Kohoutova, E.: Patient-specific fibre-based models of muscle wrapping. Interface Focus 2013 3(2), 1–8 (2013), doi:10.1098/rsfs.2012.0062
11. Marsden, S.P., Swailes, D.C.: A novel approach to the prediction of musculotendon paths. Journal of Engineering in Medicine 222(1), 51–61 (2008)
12. OpenSim project (2010), https://simtk.org/home/opensim

Long-Bone Fracture Detection in Digital X-ray Images Based on Concavity Index

Oishila Bandyopadhyay[1], Arindam Biswas[1], and Bhargab B. Bhattacharya[2]

[1] Department of Information Technology, Bengal Engineering and Science University, Howrah
[2] Center for Soft Computing Research, Indian Statistical Institute, Kolkata

Abstract. Fracture detection is a crucial part in orthopedic X-ray image analysis. Automated fracture detection for the patients of remote areas is helpful to the paramedics for early diagnosis and to start an immediate medical care. In this paper, we propose a new technique of automated fracture detection for long-bone X-ray images based on digital geometry. The method can trace the bone contour in an X-ray image and can identify the fracture locations by utilizing a novel concept of concavity index of the contour. It further uses a new concept of relaxed digital straight line (RDSS) for restoring the false contour discontinuities that may arise due to segmentation or contouring error. The proposed method eliminates the shortcomings of earlier fracture detection approaches that are based on texture analysis or use training sets. Experiments with several digital X-ray images reveal encouraging results.

Keywords: Medical imaging, Bone X-ray, Chain code, Digital straight line segment (DSS), Approximate digital straight line segment (ADSS).

1 Introduction

Fracture detection in X-ray images is an important task in emerging health-care automation systems [8], [6], [10]. Automated fracture identification from an orthopedic X-ray image needs extraction of the exact contour of the concerned bone structure. A fractured long-bone contour appears with irregular (uneven) or disconnected contour in the broken region. Bone contour discontinuity may also arise due to over-thresholding during segmentation of bone region from the surrounding flesh tissues and muscles. Long-bone fracture is a very common health problem, which needs immediate medical attention. A considerable number of men and women suffer from osteoporotic or accidental long-bone fracture everyday. Automated fracture detection can help the doctors and radiologist by screening out the obvious cases and by referring the suspicious cases to the specialists for closer examinations. Since bone fractures can occur in many ways, a single algorithm may not be suitable for analyzing the various types of fractures accurately. In the past, several approaches had been proposed by the researchers for detection of fractures in different bone regions.

Long-bone fractures usually refer to injuries in bones like humerus, radius and ulna, femur, tibia, and fibula. Each long-bone is divided into three regions

R.P. Barneva, V.E. Brimkov, and J. Šlapal (Eds.): IWCIA 2014, LNCS 8466, pp. 212–223, 2014.
© Springer International Publishing Switzerland 2014

- proximal, distal (two extremities), and diaphyseal [13]. A fracture of the diaphyseal is classified into three groups - simple, wedge, and complex. Tian et
al. [16] has implemented a femur fracture detection approach by computing the
angle between the axis of the neck of femur and the axis of the shaft. But this
kind of method can only work on severe fractures that have caused a significant change in the angle of the neck and shaft of the femur. Another femur
fracture detection approach uses contour generation and contour region filling
followed by Hough-transform, vertical integral projection, and statistical projection of differential curve to identify the fractured region [17]. Donnelley et al.
[7] have proposed a CAD system for the long-bone fracture detection which uses
scale-space approach for edge detection, parameter approximation using Hough
transform, diathesis segmentation followed by fracture detection using gradient
analysis. Classification, a frequently used data mining technique, has also been
used widely to detect the presence of fracture for the past few decades. These
systems combine various features (like shape, texture, and colour) extracted
from X-ray images and deploy machine learning algorithms to identify fractures
[11]. Several researchers have proposed texture analysis of bone structure or
bone mineral density estimation along with higher order statistical analysis for
fracture detection [5], [14], [12].

   In this paper, we have proposed a new technique based on relaxed digital
straight line segment (RDSS) to restore the contour discontinuity that may
arise during segmentation. Next, we use the corrected contour for identifying
the fracture locations using the concept of concavity index of a digital curve.
The proposed algorithm uses an entropy-based segmentation method [2] with
adaptive thresholding-based contour tracing [1] to generate the bone contour of
an X-ray image. The novelty of the technique lies in the fact that it rectifies
the false discontinuities of a bone contour and identifies the fracture region correctly. Experiments on several long-bone digital X-ray images demonstrate the
suitability of the proposed method for fracture related abnormality analysis.

2   Related Definitions

A digital image consisting of one or more objects, whose contour is formed with
fairly straight line edges, can be represented by a set of (exact or approximate)
digital straight line segments (DSS or ADSS). Such representations capture a
strong geometric property that can be used for shape abstraction of these objects [3]. In the proposed algorithm, some properties of DSS are utilized for
contour correction of a bone image. A few definitions related to this work are
given below:

Digital Curve (DC). A DC C is an ordered sequence of grid points (re-
presentable by chain codes) such that each point (excepting the first one) in
C is a neighbor of its predecessor in the sequence [15] (see 2(b), 2(c)).

Chain Code. It is used to encode a direction around the border between pixels.
If $p(i, j)$ is a grid point, then the grid point (i', j') is a neighbor of p, provided
that $max(|i - i'|, |j - j'|) = 1$. The chain code [9] of p with respect to its neighbor
grid point in C can have a value in $0, 1, 2, ..., 7$ as shown in Fig. 2 (a).

Digital Straight Line Segment (DSS). The main properties of DSS [15], [3] are:

(F1) The runs have at most two directions, differing by 45^0, and for one of them, the run length must be 1.

(F2) The runs can have only two lengths, which are consecutive integers.

(F3) One of the run lengths can occur only once at a time.

The necessary and sufficient conditions for a digital curve (DC) to be a DSS have been stated in the literature [15], [3]. In [15], it has been shown that a DC is the digitization of a straight line segment if and only if it has the chord property. A DC C has the chord property if, for every (p, q) in C, the chord pq (the line segment drawn in the real plane, joining p and q) lies near C, which, in turn, means that, for any point (x, y) of pq, there exists some point (i, j) of C such that $max(|i - x|, |j - y|) < 1$.

Relaxed Digital Straight Line Segment (RDSS). In this paper, we introduce the concept of Relaxed Digital Straight Line Segment, defined below:

RDSS inherits the basic property of the underlying DSS, i.e.,

(F1) At most two types of elements (chain code directions) can be present in a RDSS and these can differ only by unity, modulo eight.

Thus, a RDSS represents single pixel curve with very small curvature by a relaxed digital straight line. For example, in Fig. 1(b)), a portion of the curved line is approximated by two consecutive RDSS, R1 and R2.

A tighter condition leads to an earlier concept of Approximate Digital Straight Line Segment (ADSS) [3]. The main properties of ADSS are:

(F1) At most two types of elements can be present and these can differ only by unity, modulo eight.

(F2) One of the two element values always occurs singly.

(F3) Successive occurrences of the element occurring singly are as uniformly spaced as possible.

3   Proposed Algorithm

In an X-ray image, the bone parts appear along with the surrounding tissues or muscles (i.e., flesh). So bone region segmentation and bone contour generation is necessary for automated fracture identification process. The proposed method segments the bone region of input X-ray (Fig. 3(a)) image from its surrounding flesh using an entropy-standard deviation based segmentation method (Fig. 3(b)) and then applies an adaptive thresholding based technique to generate the bone contour [2][1] (Fig. 3(c)). Any discontinuity that appears in bone contour during segmentation, is corrected by the proposed method using RDSS (Fig. 3(d)). Bone fracture regions are then identified by analyzing concavity index of the corrected image (Fig. 3(e)).

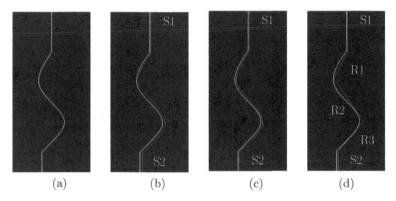

Fig. 1. RDSS (a) Straight line segment and curve, (b) corresponding DSS (35 segments), (c) corresponding ADSS (16 segments), (d) corresponding RDSS (5 segments)

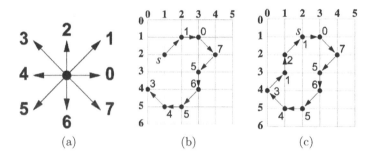

Fig. 2. Chain code (a) Chain code in 8-neighbor connectivity, (b) DC (1, 2)(10756543), (c) DC (2, 1)0756543121

3.1 Bone Region Segmentation from an X-ray Image

The major challenge in segmentation of bone region in any X-ray image lies in identification and extraction of flesh to bone transition region. Overlapping intensity range of flesh and bone region restricts the use of pixel-based thresholding or edge-based approaches as they often fail to produce accurate results. In the proposed algorithm, local entropy image is generated from the input X-ray image [2]. Local entropy image clearly identifies the flesh-bone transition points in a X-ray image with bright bone region and relatively darker flesh region. However, some X-ray images appear with bright flesh region resulting in overlapping flesh and bone intensity range. In such cases, the entropy image often fails to identify the flesh to bone transition correctly. To overcome this problem, we compute local standard deviation for each pixel and multiplied it with local entropy to facilitate bone image segmentation [1].

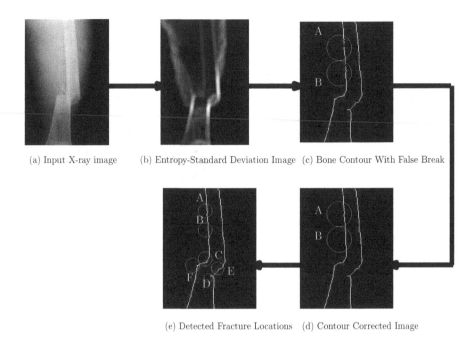

(a) Input X-ray image   (b) Entropy-Standard Deviation Image   (c) Bone Contour With False Break

(e) Detected Fracture Locations   (d) Contour Corrected Image

Fig. 3. Different phases of proposed algorithm

3.2   Contour Generation Using Adaptive Thresholding

The proposed approach generates the bone contour from a segmented bone image (entropy-standard-deviation image) using an adaptive thresholding method [1]. In adaptive thresholding based bone contour generation technique, the segmented bone image J is traversed using a small window. For each pixel α of J, the window is constructed with its 8 neighboring pixels. The window is divided into four cells, top-left (C_1), top-right (C_2), bottom-left (C_3) and bottom-right (C_4). All these four cells are incident on α. To determine whether a cell C_i has a portion of bone boundary in it, adaptive thresholding approach is used [4]. The contour traversal algorithm checks the intensity values of neighboring pixels and selects the next pixel position whenever it encounters a pixel whose intensity value exceeds the adaptive threshold value of the present pixel. After selection of a new pixel position, the algorithm checks whether or not the pixel is already visited. If the pixel is found visited, then the algorithm starts searching from a new position; otherwise it adds the current pixel in the visited list and decides the direction for the next move [1].

3.3   Contour Correction Using Relaxed Digital Straight Line Segments

A bone fracture may cause a disconnected or irregular (uneven) bone contour. Hence, any discontinuity in the bone contour that appears during segmentation

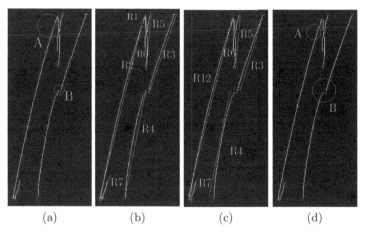

(a)	(b)	(c)	(d)

Fig. 4. Contour correction (a) Bone contour with false break, (b) Contour with RDSS, (c) Corrected contour with RDSS, and (d) Corrected contour

or contour generation, can mislead fracture detection . To overcome this problem, we propose a novel approach based on relaxed digital straight line segment (RDSS).

In the proposed method, the bone contour generated using adaptive thresholding approach is traversed from top-left to bottom and from bottom-right to top and the corresponding chain code list is generated. This chain code list is analyzed to approximate the underlying curve with relaxed digital straight line segments (RDSS). If any discontinuity arises in bone contour during segmentation or contour generation process (see region A of Fig. 4(b)), then it should have been be covered by two different RDSS. The proposed algorithm searches for all such RDSS pairs that cover the bone contour with same or its complementary chain code string. For example, a RDSS with chain code consisting of {1,2} can be paired with another having the code {5,6} as the RDSS covering the two line segments across the break will be traversed either from the same (or opposite) direction, with the start and end pixels lying in close neighborhood of each other. After finding such RDSS pairs, any one of these two RDSS is extended in the direction of traversal to connect them into a single RDSS (see Fig. 4(c). Any discontinuity in the bone contour caused by a fracture (see region B of Fig. 4(d)) will change their alignment. Therefore, a RDSS cover of such discontinuities cannot be extended to combine them into a single RDSS. On the other hand, the contour discontinuities caused by segmentation or contouring errors can be corrected by projecting the two neighboring RDSS towards each other.

Approximation of a bone contour can also be performed using DSS or ADSS; however, the number of segments required to cover the contour would be very high as in a single-pixel wide long-bone image, the contour usually changes the direction at an interval of every 5 to 10 pixels. Thus the use of RDSS not only reduces the number of approximating straight line segments, but also rectifies the false breaks in the contour, while reporting the correct fracture locations (see Fig. 4(d)).

3.4   Fracture Detection Using Concavity Index

In the proposed algorithm, we introduce the concept of concavity index, which is used to detect the fractures in long-bones.

Concavity Index (α). During the traversal, each point p_i is assigned a concavity index, α_i where α_0 and α_1 are initialized to 0 and 1 respectively. To obtain α_{i+1}, α_i is incremented (decremented) by the difference of the directions, d_i and d_{i+1}, if the contour moves in clockwise (counter-clockwise) direction from p_i to p_{i+1}, where d_i is the incident direction at p_i.

It should be noted that if the curve propagates in the same direction, the concavity index remains unchanged, however, a significant variation of α indicates abnormalities in the curve leading to fracture detection.

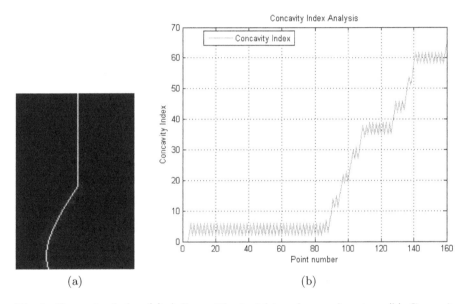

(a)   (b)

Fig. 5. Concavity Index (a) A line with straight and curved region, (b) Concavity indices of different points of the curve shown in (a)

In the above example (Fig. 5), it is clearly shown that concavity index changes significantly in a curved region. The proposed method uses this property of concavity index in fracture detection of long-bone contour.

We traverse the bone contour chain code and computes the concavity index for all pixel positions during traversal. We plot concavity index against the pixel positions as we traverse the contour. In a fractured long-bone, the bone boundary appears as a curve in the fractured part of the contour as the abnormality therein degrades its straightness. The proposed algorithm thus locates the fractured regions by observing the sharp and frequent changes (wave-like structure with peak and fall) in concavity index value (see Fig. 6(b)). Experimental evidences show that

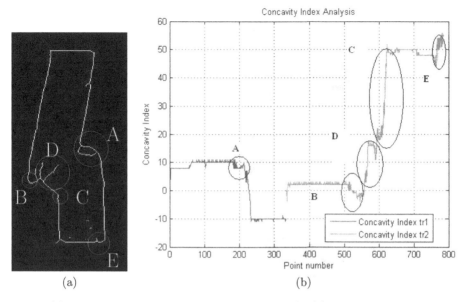

(a)   (b)

Fig. 6. (a) Contour of input image with fracture marked (b) Concavity index plot for
each traversal of (a) with marked changes in fracture location.

in the fractured regions, the change in chain code is more than 3 and change in concavity index exceeds 10 over a contour length of around 5 or 6 pixels. The groups
of neighbouring pixels identified with such values are then clustered to identify the
fracture zones marked in the concavity index curve (Fig. 6(b)). In this example,
regions marked as 'A', 'B', 'C', 'D' identify the fractured region correctly and that
marked as 'E' shows wrong identification (false positive).

4   Experimental Results

The proposed method is tested on several X-ray images of long-bone fracture.
For each case, the concavity index curve is analyzed to identify the fractured
locations. It is noticed that for most of the cases the fractured locations are
identified correctly. Concavity index is computed during bone contour tracing
from top-left corner to bottom and from bottom-right corner to top (plot with
different colour in 7). A plot of concavity index against the traversed pixel list
shows that the concavity index increases gradually (region 'A' in concavity index
curve of Fig. 7(h)) during the traversal of regular bone curvature (as shown in
region 'A' of Fig. 7(g)). A traversal of the fractured regions shows a fast change
in te concavity index; this generates a wave like structure with peaks and falls in
the curve (see region marked with 'A', 'B', 'C', 'D', 'E' of Fig 7(c)) and respective
regions in concavity index curve (see region marked with 'A', 'B', 'C', 'D', 'E'
of Fig 7(d)). Table -1 shows the concavity curve analysis for each X-ray image.
It is noticed that the number of false positives identified in the fractured region

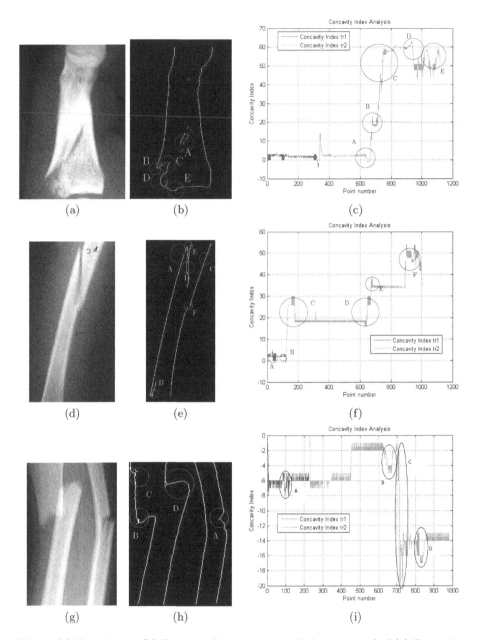

Fig. 7. (a) X-ray image (b) Contour of input image with fracture marked (c) Concavity index plot for each traversal of (a) with marked changes in fracture location; (d) X-ray image (e) Contour of input image with fracture marked (f) Concavity index plot for each traversal of (d) with marked changes in fracture location; (g) X-ray image (h) Contour of input image with fracture marked (i) Concavity index plot for each traversal of (g) with marked changes in fracture location

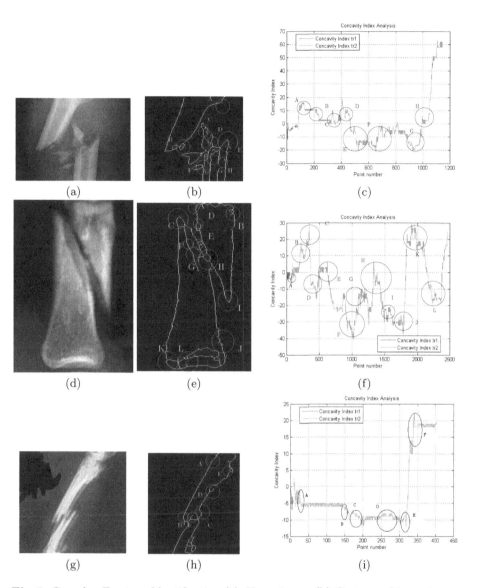

Fig. 8. Complex Fracture Identification (a) X-ray image (b) Contour of input image with fracture marked (c) Concavity index plot for each traversal of (a) with marked changes in fracture location; (d) X-ray image (e) Contour of input image with fracture marked (f) Concavity index plot for each traversal of (d) with marked changes in fracture location; (g) X-ray image (h) Contour of input image with fracture marked (i) Concavity index plot for each traversal of (g) with marked changes in fracture location

Table 1. Concavity Index Analysis for Fracture Detection in X-Ray Images

Input image	Fracture region identified	Correct identification	False positive
Fig. 6(a)	5	4 (A, B, C, D)	1 (E)
Fig. 7(b)	4	4 (B, C, D, E)	0
Fig. 7(e)	7	4 (A, D, E, F)	2 (C, B)
Fig. 7(h)	4	3 (A, B, D)	1 (C)
Fig. 8(b)	8	6 (B, C, D, E, F)	2 (A, H)
Fig. 8(e)	12	8 (A, C, D, E, F, G, H, I)	4 (B, J, K, L)
Fig. 8(h)	6	3 (B, C, F)	3 (A, D, E)

of each image is considerably low. Fig. 8 shows some example of more complex fractures where bones are fragmented into multiple pieces. The concavity curve of these X-ray images appears as multiple waves (see Fig. 8(c), Fig. 8(f), Fig. 8(i)), which clearly indicate the presence of fractures in the input image.

5   Conclusion

We have proposed a method for fracture detection in X-ray images based on digital geometry. We have shown, for the first time, that the power of digital geometric techniques can be harnessed to provide fast and accurate solutions to the automation of medical image analysis. Our experiments on several X-ray image databases demonstrate its suitability of fracture detection in long-bone structures.

References

1. Bandyopadhyay, O., Biswas, A., Chanda, B., Bhattacharya, B.B.: Bone contour tracing in digital X-ray images based on adaptive thresholding. In: Maji, P., Ghosh, A., Murty, M.N., Ghosh, K., Pal, S.K. (eds.) PReMI 2013. LNCS, vol. 8251, pp. 465–473. Springer, Heidelberg (2013)
2. Bandyopadhyay, O., Chanda, B., Bhattacharya, B.B.: Entropy-based automatic segmentation of bones in digital X-ray images. In: Kuznetsov, S.O., Mandal, D.P., Kundu, M.K., Pal, S.K. (eds.) PReMI 2011. LNCS, vol. 6744, pp. 122–129. Springer, Heidelberg (2011)
3. Bhowmick, P., Bhattacharya, B.B.: Fast polygonal approximation of digital curves using relaxed straightness properties. IEEE Transactions on Pattern Analysis and Machine Intelligence, 1590–1602 (2007)

4. Biswas, A., Khara, S., Bhowmik, P., Bhattacharya, B.B.: Extraction of region of interest from face images using cellular analysis. ACM Compute 2008, 1–8 (2008)
5. Chai, H.Y., Wee, L.K., Swee, T.T., Salleh, S.H., Ariff, A.K., Kamarulafizam: Gray-level co-occurrence matrix bone fracture detection. American Journal of Applied Sciences, 26–32 (2011)
6. Donnelley, M., Knowles, G.: Automated bone fracture detection. In: Proceedings of SPIE 5747, Medical Imaging: Image Processing, p. 955 (2005)
7. Donnelley, M., Knowles, G., Hearn, T.: A CAD system for long-bone segmentation and fracture detection. In: Elmoataz, A., Lezoray, O., Nouboud, F., Mammass, D. (eds.) ICISP 2008 2008. LNCS, vol. 5099, pp. 153–162. Springer, Heidelberg (2008)
8. Eksi, Z., Dandil, E., Cakiroglu, M.: Computer-aided bone fracture detection. In: Proceedings of Signal Processing and Communications Applications, pp. 1–4 (2012)
9. Freeman, H.: On the encoding of arbitrary geometric configurations. IRE Trans. Electronic Computers, 260–268 (1961)
10. Hacihaliloglu, I., Abugharbieh, R., Hodgson, A.J., Rohling, R.N., Guy, P.: Automatic bone localization and fracture detection from volumetric ultrasound images using 3-d local phase features. Ultrasound Med. Biol. (1), 128–144 (2012)
11. Lum, V.L.F., Leow, W.K., Chen, Y.: Combining classifiers for bone fracture detection in X-ray images. In: IEEE International Congress on Image and Signal Processing, 1149–1152 (2005)
12. Materka, A., Cichy, P., Tuliszkiewicz, J.: Texture analysis of X-ray images for detection of changes in bone mass and structure. In: Texture Analysis in Machine Vision. p. 257, World Scientific (2000)
13. Muller, M.E., Nazarian, S., Koch, P., Schatzker, J.: The comprehensive classification of fractures of long bones. Springer (1990)
14. Ouyang, X., Majumdar, S., Link, T.M., Lu, Y., Augat, P., Lin, J., Newitt, D., Genant, H.K.: Morphometric texture analysis of spinal trabecular bone structure assessed using orthogonal radiographic projections. Medical Physics Research and Practice, 2037–2945 (1998)
15. Rosenfeld, A.: Digital straight line segments. IEEE Transactions on Computers, 1264–1269 (1974)
16. Tian, T.-P., Chen, Y., Leow, W.-K., Hsu, W., Howe, T.S., Png, M.A.: Computing neck-shaft angle of femur for X-ray fracture detection. In: Petkov, N., Westenberg, M.A. (eds.) CAIP 2003. LNCS, vol. 2756, pp. 82–89. Springer, Heidelberg (2003)
17. Wei, Z., Liming, Z.: Study on recognition of the fracture injure site based on X-ray images. In: IEEE International Congress on Image and Signal Processing, pp. 1947–1950 (2010)

Smoothing Filters in the DART Algorithm

Antal Nagy

Department of Image Processing and Computer Graphics, University of Szeged
nagya@inf.u-szeged.hu

Abstract. We propose new variants of the Discrete Algebraic Reconstruction Technique (DART) with a combined filtering technique. We also set up a test framework to investigate the influence of the filters for different number of sources and noise level in case of various parameters. Our results are produced by performing numerous reconstructions on the test data set. The reconstructed images were evaluated by locally using relatives mean error (RME) and globally by an ordered ranking system. The achievements are subjected and discussed. Finally we also suggest a filter parameter combination which gives a way to improve the quality of the DART reconstruction algorithm.

Keywords: Discrete tomography, Reconstruction quality, DART, Filtering, Non-destructive testing.

1   Introduction

The aim of the *tomography* is to reconstruct cross-sections of the object studied which is determined by projection images. The *Discrete Tomography* (DT) can be applied when the object consists of only few homogeneous materials with known attenuation coefficients (see in [9]). *Binary tomography* is a special case of DT where we have only e.g. the object and the air to detect from the projections.

Several reconstruction methods are proposed to reconstruct discrete objects which are based on energy minimization problem (see in [14] and [12]), using statistical model (e.g. in [1]), and modeling the reconstruction problem as a graph network flow problem (see in [2]).

DART reconstruction algorithm [3] is an iterative, heuristic reconstruction algorithm where the continuous reconstructions are thresholded within each iteration step. After the thresholding the DART modifies the boundary pixels by an other continuous reconstruction on the boundary pixels only. These pixels are independently changing, so a *smoothing operation* is applied necessarily as a final action of the iteration.

Many succeeding versions of the DART algorithm are published (e.g. [4] and [11]) which shows that this method is really in the focus and can be used in practical cases. Previously reconstruction enhancement was studied in [7] for Algebraic Reconstruction Methods (ARM) and the DART algorithm with a specific median filter.

In this paper we deal with the experimental investigation of the smoothing filters' behavior during the DART iterations. The method for our study is a

R.P. Barneva, V.E. Brimkov, and J. Šlapal (Eds.): IWCIA 2014, LNCS 8466, pp. 224–237, 2014.

simulation framework where we observe the quality of the reconstruction algorithm. We give a comprehensive investigation of different filters and parameter combinations.

The structure of the paper is as follows. In the first section the reconstruction problem is introduced. After that we shortly present the smoothing filters. In the Section 4 we propose new variants of the DART reconstruction algorithm. In the next two sections we review the measurements we used in our experimental tests and the experimental setup as well. The results and the discussions are detailed in the Section 7. Finally, the last chapter gives a conclusions we obtained and the possible further works.

2   The Reconstruction Problem

In our experimental study we used parallel beam geometry for the two dimensional transmission tomography which is a frequently used geometry setup e.g. in synchrotron transmission tomography.

The binary reconstruction problem from parallel beam geometry can be reformulated as linear equation system

$$\mathbf{A}\mathbf{x} = \mathbf{b}, \tag{1}$$

where $\mathbf{A}$ matrix contains the projection geometry, $\mathbf{b}$ denotes the measured projection vector, and $\mathbf{x} \in \{0,1\}^{n^2}$ contains the unknown image pixels to be reconstructed with size of n^2. The task is to determine the $\mathbf{x}$ unknown image from (1).

The equation system can be under determined and the reconstruction problem has more than one solution when we have few projections. Moreover due to the measured projection errors it is possible that there is no solution of the (1) at all.

3   Smoothing Filters

Different kinds of techniques are used to reduce noise effects for continuous reconstruction algorithms. Some methods suggest pre-filtering of the projection data. E.g. in the Filtered Backprojection continuous reconstruction method, convolution is performed with a selected kernel in order to avoid the aliasing problems during the back projection. Some iterative ARM algorithms (e.g. SIRT in [8])) have some smoothing effects, too. Also post reconstruction techniques exist to remove the artefacts caused by distortions.

In the DART algorithm a smoothing operation is required due to the severally altering boundary pixels. According to the results of the [7] we have performed preliminary studies to investigate the possible smoothing filters which can be used in DART algorithm.

Due to the fact that the filtering is executed in every DART iterations, in that way it has serious impact for the computational complexity. In hence e.g. the adaptive version of the median filtering would consequentially raise the running time of the modified DART algorithm (see Section 4) likewise all the iterative noise reduction methods.

We also had to taken into account that the noise filtering methods have various parameter lists. These parameters also influence the quality of the reconstruction and usually takes lots of time to find the optimal settings as it we have found in our forehand tests, namely the quality of the reconstructions was not significantly improved for the selected parameters using other filters in the DART.

In that way we selected only those filters where the parameter lists were comparable with the preliminarily used filters and fitted in our complex framework (see Section 6) to find the best parameter combination and relatively would not increase the computational time of the modified DART algorithm:

Median filter was investigated previously in [7]. This filter is generally used to remove the raised values from a certain environment, meanwhile it keeps the structures of the given image during the enhancement procedure.

Gauss filter was also used in DART before in [4] and [13] to reduce the noise. Gaussian filter is also a well known filter used in image enhancement which convolves the values with a Gaussian function.

Bilateral filter is the third one selected which is a non-linear edge-preserving filter. During the filtering the values are replaced by the weighted average surrounding pixels. The weights are based on Gaussian distribution but it takes into account the large differences between the pixel values which reduce the effect of the smoothing at the edges. It was introduced in [15] and many accelerated versions of the given filter were published (e.g. in [5] and [6]) previously.

4   Modified DART Algorithm

The number of the sources and the number of the iteration steps can be reduced by using median filter in the DART algorithm to get the satisfying reconstruction quality shown in [7]. The problem was that the optimal smoothing filter parameter was highly depending on the object structure to be reconstructed.

We have modified the original DART algorithm as it can be seen in Algorithm 1 to incorporate the previously shown filters.

In Algorithm 1 the alternative smoothing operators can be given as $S_{\text{Alt}} \in \{M_w, G_w, B_w\}$, where the capital letters are coming from the first letter of the filter's name (Median, Gauss, and Bilateral respectively) and the w index denotes the window size of the filter.

The t_s input parameter of the Algorithm 1 gives the frequency of the usage of the filter during the reconstruction iterations. In case $t_s < 0$ only the alternative smoothing operator is used during the reconstruction. Otherwise, in every $t_s{}^{\text{th}}$ iteration step the S_{Alt} is applied before the S_{orig} smoothing filter. S_{orig} was chosen according to [3] where pixels in set B^k were convolved by H kernel (see in (2)):

Algorithm 1. Discrete Algebraic Reconstruction Technique using different filters

Input: A projection matrix; b expected projection values; κ threshold function, S_{Alt} alternative smoothing operator, t_s smoothing frequency; S_{orig} original smoothing operator

1: Compute a starting reconstruction $\mathbf{x}^{(0)}$ using an algebraic reconstruction method
2: $k \leftarrow 0$
3: **repeat**
4:    $k \leftarrow k + 1$
5:    Compute a thresholded image $\mathbf{r}^{(k)} = \kappa(\mathbf{x}^{(k-1)})$
6:    Compute $I^{(k)}$ set of non-boundary pixels of $\mathbf{r}^{(k)}$
7:    Compute $J^{(k)}$ set of boundary pixels of $\mathbf{r}^{(k)}$
8:    Compute the $\mathbf{y}^{(k)}$ in the following way $y_i^{(k)} \leftarrow \begin{cases} r_i^{(k)}, & \text{if } i \in I^{(k)}, \\ x_i^{(k-1)}, & \text{if } i \in J^{(k)}. \end{cases}$
9:    Using $\mathbf{y}^{(k)}$ as a starting solution, compute a continuous reconstruction $\mathbf{x}^{(k)}$ while keeping the pixels in $I^{(k)}$ fixed
10:   **if** $t_s < 0$ **then**
11:      Apply S_{Alt} a smoothing operation to the pixels that are in $J^{(k)}$
12:   **else**
13:      **if** $(t_s \bmod k) == 0$ **then**
14:         Apply S_{Alt} a smoothing operation to the pixels that are in $J^{(k)}$
15:      **end if**
16:      Apply S_{orig} smoothing operation to the pixels that are in $J^{(k)}$
17:   **end if**
18: **until** a stopping criteria is met.
19: **return** the segmented image $\kappa(\mathbf{x}^{(k)})$

$$H = \begin{pmatrix} 3/80 & 3/80 & 3/80 \\ 3/80 & 7/10 & 3/80 \\ 3/80 & 3/80 & 3/80 \end{pmatrix}. \tag{2}$$

The DART variant Algorithm 1 combines the S_{orig} and the S_{Alt} filters.

The κ threshold function for binary tomography (i.e. $\mathbf{x} \in \{0,1\}^{n^2}$) was defined in the following way:

$$\kappa(v) = \begin{cases} 0 & (v < 0.5), \\ 1 & (v \geq 0.5). \end{cases} \tag{3}$$

Defining all the parameters of the Algorithm 1 besides the original arguments, we can construct the set of new variants of the DART algorithm. We should mention that Gauss and Bilateral filters have additional parameters which influence the behavior of the new version of DART algorithm as well.

In the next chapters we present the experimental tests were performed on the given dataset.

5 Measurements

In order to measure the quality of the results we compared them to the original phantom images using Relative Mean Error (RME) :

$$\text{RME} = \frac{\sum\limits_{j=1}^{J} |x_j - \hat{x}_j|}{\sum\limits_{j=1}^{J} \hat{x}_j} \cdot 100\% \,, \tag{4}$$

where $\hat{\mathbf{x}} = \{\hat{x}_j\}_{j=1}^{J}$ denotes the vector of the original phantom image (see in [10]). Clearly, RME ≥ 0 and the smaller value indicates a better comparison result. Furthermore, RME $= 0$ if and only if $\mathbf{x} = \hat{\mathbf{x}}$.

The RME can only be considered as a measure of a quality of the reconstruction when we are reconstructing one phantom from a given number of projections which is degraded with given noise values added to the projections. Comparing only the RME values of the reconstructions' results of different phantoms would give us false outcomes because it measures only the distances from the exact solutions which can be very discrepant using a bunch of reconstruction algorithms on various projections of diverse phantoms.

For that reason mentioned above, we have introduced the following ranking system. Accordingly to the actual parameter settings, we have performed DART reconstructions with different filters on a phantom's projections. We calculated the RME values separately for each reconstruction results. After that we put the RME values algorithms in a row. It gives us an order of the reconstruction algorithms used in the given setup. Summarizing these places we could have higher level view of the behaviors of the examined reconstruction algorithms. For example summarizing the rankings for all test phantoms when number of sources is 3 and the noise level is 0% one can decide which algorithm outperforms the others for all phantoms with the presented setup. Using this aggregation we are able to calculate a global order of the investigated reconstruction algorithms as well.

6 Experimental Setup

In this section we describe the test environment of our experimental investigation of the set of DART variants and the basic DART (see in [3]) reconstruction algorithms.

The Test Dataset

We have chosen 21 binary software phantoms with size of 256 by 256 pixels. One part of the binary images are from the image database of the IAPR Technical Committee on Discrete Geometry (TC18), and the remaining phantoms previously were published (e.g. in [14], [12], and [3]). The selected 21 binary images provide us high variety of shapes to perform our comprehensive experiments. Some example of the phantoms are presented in Fig. 1.

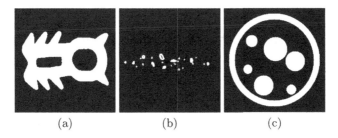

(a)   (b)   (c)

Fig. 1. Some example of the software phantoms used in our experiments

Projection Geometry Setup

In our implementation, the projections were created according to (1) where $\mathbf{A}$ contains the real weights of the area to be reconstructed in contrast with execution in [7], where the projection system matrix contains only binary values. Our implementation gives better approximation of the projection values, in hence the reconstruction results and the measurements of the reconstruction quality would be more realistic.

We denote the number of sources of the given projection geometry with K where $K \in [3, 5, 10, 15]$, where the sources were placed uniformly from $0°$ to $180°$. These K values are generally used in binary tomography problems where we assume that the investigated object made of only one material.

Noise Parameter

To simulate the measured error we used additive Gaussian noise on the projection data which is usually acceptable in transmission tomography. The noise contributions were denoted by η and the values were chosen from noiseless case to higher noise level ($\eta \in [0\%, 1\%, 3\%, 5\%, 10\%]$). In our experiments at each noise level a random Gaussian noise with size of the projection was generated with zero mean value and with the given η standard deviation and added to the exact projection.

Filter Parameters

The selected filters have several parameters. During the experiments the similar parameters were set in the same way in order to make the filter properties in consistent form. For those parameters where we could not set them in the same way, we selected one fix parameter for the experimental tests.

All filters in our selection have w window size parameter where it was adjusted in the following way: $w \in [3, 5, 7, 9]$.

In case of the spatial Gaussian σ_S parameter of the Gauss and Bilateral filter, the w window size were used.

Due to the binary case of reconstruction problem the with of the range Gaussian parameter was set to 2 of the Bilateral filter. The truncation error parameter of the implementation (see in [5] and [6]) was set to 1.75 empirically.

Usage Frequency

In Algorithm 1 the t_s smooth frequency controls how the S_{Alt} alternative smoothing operator was applied during the DART iterations. The values were selected from the given set $t_s \in [-1, 1, 2, 3, 5]$, where -1 denotes that only the alternative filter was used in the iterations.

DART Parameters

During the experimental tests we used the same parameters for all DART variants and the original DART as well. We have performed 10 SIRT as an Algebraic Reconstruction Method (ARM) iterations for all 100 DART iterations in the reconstruction process.

7   Results and Discussion

We are going to present our results as it was introduced in the Section 5. We will denote the DART variants with the filter name, window size, and the usage frequency parameters were used during the reconstructions respectively. For example the notation Median_7_3 means that the median filter was used with size of 7 in every third DART iteration as the S_{Alt} before the S_{Orig} filtering. Gauss_9_-1 means that only the Gaussian filter was used with 9 window size parameter in every iterations. The original DART was named as DART_ in our results. In the tables we present the top 5 and last 5 places. The last 5 ranks is denoted from -5 to -1 where -1 is the last in the ordered list. The last column represents the aggregated results for all noise levels for the selected number of sources.

We have selected the representative reconstruction results with their RME values and the reconstructed images to show the differences between the first top 5 and the last 5 sections. These representatives are marked with *slanted* fonts in the Tables. We left out from the representative results the original DART_ algorithm because the S_{Orig} kernel has a relatively small w windows size comparing to other variants and in most of the cases the DART variants use other filters together with the S_{Orig}.

All reconstruction results will be demonstrated on Fig. 1.a phantom. The first image row shows the selected top 5, the second shows the selected last five representative results.

In Table 1 we show the results summarized for all reconstructed phantoms where the number of sources was equal to 3. We can see that in the top 5 mostly the Gaussian DART variants can be found with larger window size ($w \gtrsim 7$) and smaller frequency values ($t_s \lesssim 3$).

Table 1. The order of the algorithms where $K = 3$ for all phantoms

Rank \ Noise	0%	3%	5%	10%	All noise
1	Gauss_7_-1	Bilat_9_-1	Bilat_7_3	Gauss_9_3	Gauss_9_3
2	Gauss_5_-1	Gauss_9_-1	Median_9_3	Median_7_3	Gauss_9_-1
3	Gauss_7_3	Gauss_9_1	Gauss_9_-1	Median_9_3	Bilat_9_-1
4	Bilat_9_1	Gauss_7_1	Median_7_3	Bilat_7_1	Gauss_7_3
5	Gauss_9_5	Gauss_9_3	Gauss_9_3	Bilat_7_-1	Gauss_7_-1
-5	DART__	Gauss_3_5	Bilat_3_2	Gauss_3_2	Bilat_3_2
-4	Bilat_5_-1	Median_3_5	Gauss_3_5	Gauss_3_5	Gauss_3_5
-3	Bilat_7_-1	Bilat_3_2	Bilat_3_5	Median_3_5	Median_3_5
-2	Median_3_-1	Bilat_3_5	DART__	Bilat_3_5	Bilat_3_5
-1	Bilat_3_-1	DART__	Median_3_5	DART__	DART__

Also we can observe that in the last 5 rank section the frequency $t_s \simeq 5$ parameter values were the typical cases with smaller window size $w \simeq 3$. In noiseless case the low t_s values are more typical.

These observation subsequent upon the low number of sources. The reconstruction problem is extremely under determined in case of $K = 3$ (see the first column of the Fig. 5). Using larger w window size parameter for filters we are able to interpolate the missing information up to a certain limit. Combining large w window size with low t_s parameter at $\eta = 0\%$ level we over-smooth the object to be reconstructed in the last 5 section. At higher noise level the larger t_s parameter is not enough to correct the projection distortion caused by additive noise during the DART iterations.

Due to the low number of sources ($K = 3$) we got results with high RME values (see examples the first column of Fig. 5). For that reason we do not give here a detailed representative examples.

When $K = 5$ (see Table 2) the filters with larger window size and at lower t_s values can be found in the top 5 again. While the lower window size and larger frequency parameter values are the typical filters except in case when $\eta = 0\%$ in the last 5 section. In noiseless instance the large window size with lower t_s parameter gives result with lower quality for the $(-5 \cdots - 1)$ places.

The explanation can be similar to the previous observation where the filter size and frequency parameter combination was depending on the number of sources and the noise level.

The Fig. 2 shows the selected representative results of the Table 2. We can see that at lower level of noise we can obtain much better result comparing the RME values. When $\eta = 0\%$ we got 13 times better RME value using Median_7_5 filter compared to Median_9_1 filter performance reconstructing Fig. 1.a phantom. At $\eta = 10\%$ noise level the usage of Gauss_9_1 filter still outperforms the Bilat_3_5 filter using it in the Algorithm 1 DART variant algorithm.

When $\eta = 0\%$ the Bilateral filter with smaller window size and $t_s \gtrsim 2$ can be found in the first 5 places in case of $K = 10$ (see Table 3). At higher level of

Table 2. The order of the algorithms where $K = 5$ for all phantoms

Rank \ Noise	0%	3%	5%	10%	All noise
1	Median_7_5	Gauss_7_1	Gauss_9_-1	Gauss_9_1	Bilat_7_1
2	Median_5_5	Bilat_9_3	Bilat_9_-1	Gauss_9_-1	Bilat_9_1
3	Median_3_1	Gauss_9_-1	Gauss_9_1	Bilat_9_1	Gauss_9_-1
4	Median_5_2	Gauss_9_3	Bilat_7_1	Median_7_1	Gauss_9_3
5	Median_3_2	Bilat_7_-1	Median_9_3	Median_7_-1	Bilat_9_3
-5	Bilat_9_-1	Gauss_3_2	Bilat_3_2	Bilat_5_5	Bilat_3_2
-4	Gauss_9_1	Bilat_3_2	Gauss_3_5	Gauss_3_5	Median_3_5
-3	Median_9_-1	Gauss_3_5	Bilat_3_5	Median_3_5	Gauss_3_5
-2	Bilat_5_-1	Bilat_3_5	Median_3_5	Bilat_3_5	Bilat_3_5
-1	Median_9_1	DART__	DART__	DART__	DART__

Table 3. The order of the algorithms where $K = 10$ for all phantoms

Rank \ Noise	0%	3%	5%	10%	All noise
1	Bilat_3_1	Bilat_7_3	Bilat_9_3	Gauss_9_1	Bilat_7_1
2	Bilat_3_2	Gauss_7_3	Bilat_7_1	Bilat_9_1	Bilat_9_3
3	Bilat_3_5	Gauss_5_1	Gauss_7_1	Gauss_9_3	Bilat_7_3
4	Bilat_3_3	Bilat_9_3	Gauss_9_3	Gauss_9_-1	Bilat_9_1
5	Bilat_5_2	Gauss_5_-1	Gauss_7_-1	Gauss_7_1	Bilat_5_1
-5	Median_9_3	Gauss_5_5	Bilat_5_5	Bilat_5_5	Bilat_3_5
-4	Gauss_9_-1	Bilat_3_5	Gauss_3_5	Gauss_3_5	Gauss_5_5
-3	Gauss_9_1	Median_3_5	Median_3_5	Median_3_5	Median_3_5
-2	Median_9_1	Gauss_3_5	Bilat_3_5	Bilat_3_5	Gauss_3_5
-1	Median_9_-1	DART__	DART__	DART__	DART__

noise the Gaussian is the bulk of filters in the top 5 section with larger window size and lower t_s value.

The characteristic w window and t_s frequency parameter values in the last 5 places can be found in noiseless and noisy cases as well. In case of $\eta \neq 0\%$ the t_s values are equal to 5 in the last section with relatively small window size. In noiseless case the large window size and small t_s values gave worse results.

$K = 10$ is treated as relatively high number of sources in DT therefore we can say that the observed parameter filters are depending on the noise level only in that case.

The Fig. 3 presents the selected representatives of the Fig. 1a of the Table 3. The RME values show high differences between the upper and lower sections. In the first column of the Fig. 3 using the Bilat_3_1 filter gives almost 20 times better RME value than the Median_9_-1 filter. When $\eta = 10\%$ the Gauss_9_1 gives better result than the Bilat_3_5 filter also. The differences also can be demonstrated with the reconstructed images in the last column of the Fig. 3.

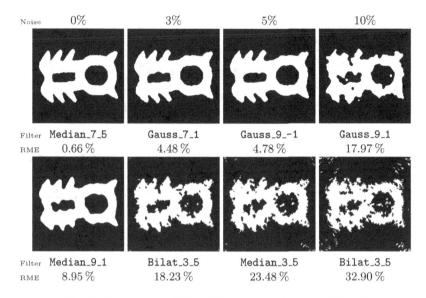

Noise	0%	3%	5%	10%
Filter	Median_7_5	Gauss_7_1	Gauss_9_-1	Gauss_9_1
RME	0.66 %	4.48 %	4.78 %	17.97 %
Filter	Median_9_1	Bilat_3_5	Median_3_5	Bilat_3_5
RME	8.95 %	18.23 %	23.48 %	32.90 %

Fig. 2. Reconstructed Fig. 1(a) phantom in case of $K = 5$

The last group of our observations when the number of sources is 15 (see Table 4). In the first 5 places taking into account the increasing level of the noise the larger window size and smaller t_s values gave better results. We can also notice that at the lower level of noise the Bilateral is the typical filter in the top 5 section while at the higher level of noise the Gaussian filter supersedes the Bilateral filter. For the filter parameters the opposite tendency can be seen in the last five places.

Considering the filter type we can not determine any correlation with the increasing amount of noise. Here we can say that the applied filters serve the noise reduction during the reconstructions.

Table 4. The order of the algorithms where $K = 15$ for all phantoms

Rank \ Noise	0%	3%	5%	10%	All noise
1	Bilat_3_1	Gauss_5_1	Bilat_7_1	Gauss_9_1	Bilat_7_1
2	Bilat_3_2	Gauss_5_3	Bilat_9_3	Bilat_9_1	Bilat_9_3
3	Bilat_3_3	Gauss_5_-1	Gauss_7_1	Gauss_7_1	Bilat_7_3
4	Bilat_3_5	Gauss_7_3	Bilat_7_-1	Gauss_9_3	Bilat_5_1
5	Bilat_5_1	Bilat_7_3	Bilat_7_3	Bilat_7_1	Bilat_5_3
-5	Median_9_3	Gauss_9_5	Bilat_5_5	Bilat_5_5	Gauss_9_5
-4	Gauss_9_-1	Median_3_5	Gauss_3_5	Gauss_3_5	Gauss_5_5
-3	Gauss_9_1	Bilat_3_5	Bilat_3_5	Bilat_3_5	Median_3_5
-2	Median_9_1	Gauss_3_5	Median_3_5	Median_3_5	Gauss_3_5
-1	Median_9_-1	DART__	DART__	DART__	DART__

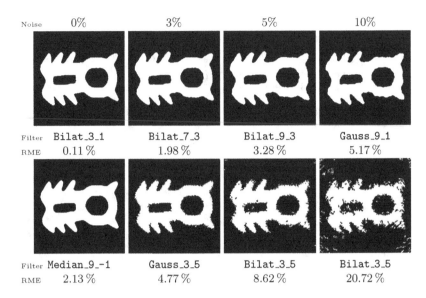

Fig. 3. Reconstructed Fig. 1(a) phantom in case of $K = 10$

In case of $K = 15$ the RME values shows that the projections provide sufficient information to reconstruct the given object with high quality (see Fig. 4). Even in noisy case we can also reach better results comparing the RME values and the reconstructed images as well.

Aggregating our previous results (see Table 5) we can present the global achievement for all number of sources. It is perceptible that the same statements can be done for the summarized ranks. We can give the best filter combination when we considering all noise levels shown in last column of the Table 5.

We present the selected global filter combinations in Fig. 5 from the Table 5 to show the differences taking into account the number of sources at 5% noise level.

Table 5. The order of the algorithms for all number of sources for all phantoms

Noise / Rank	0%	3%	5%	10%	All noise
1	Bilat_3_1	Gauss_7_3	Gauss_9_3	Gauss_9_1	*Bilat_7_1*
2	Bilat_5_1	Bilat_9_3	Gauss_7_1	Gauss_9_3	Bilat_9_3
3	Bilat_3_2	Gauss_9_3	Bilat_7_1	Bilat_9_1	Bilat_9_1
4	Median_5_5	Gauss_7_1	Bilat_9_3	Gauss_9_-1	Bilat_7_3
5	Bilat_3_5	Gauss_5_1	Gauss_9_-1	Bilat_9_-1	Gauss_9_3
-5	Gauss_9_-1	Bilat_3_2	Bilat_3_2	Bilat_5_5	Gauss_3_2
-4	Median_9_3	Median_3_5	Gauss_3_5	Gauss_3_5	Bilat_3_5
-3	Median_9_-1	Gauss_3_5	Bilat_3_5	Median_3_5	Median_3_5
-2	Gauss_9_1	Bilat_3_5	Median_3_5	Bilat_3_5	*Gauss_3_5*
-1	Median_9_1	DART__	DART__	DART__	DART__

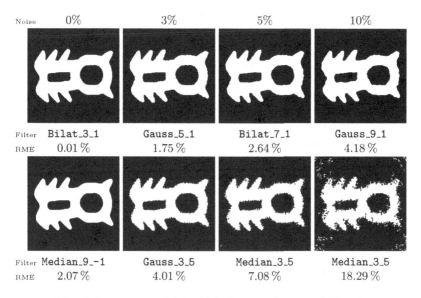

Fig. 4. Reconstructed Fig. 1(a) phantom in case of $K = 15$

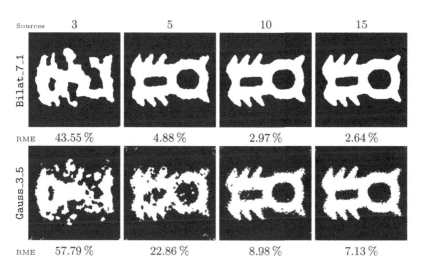

Fig. 5. Reconstructed Fig. 1(a) phantom at $\eta = 5\%$ noise level with the selected filter in Table 5

Modest improvement was gained with the selected filter from the aggregated ranking list (see Fig. 5) but we can say that Bilat_7_1 filter combination every times gave better results than the Gauss_3_5.

We also observed that the combined filters in the modified DART algorithm did not cause extraordinary computational time during the reconstructions which is not detailed in this paper.

8 Conclusions and Further Work

We have performed a comprehensive test on 21 binary software phantoms and our results were evaluated by a special ranking system. These presented assessments showed characteristic tendencies. We can conclude that the usage of the Algorithm 1 with the given filters and parameters in noisy case give us a possibility to find a combination which always gave better result in our experiments for the binary test phantoms. When no noise was added to the projections the filters with smaller window size and larger frequency values showed better performance. Increasing the noise level the larger window size and the lower t_s values were more preferable. We can also noticed that the median filter just in small number of cases produced better results.

For real applications (e.g in non-destructive testing) the DART algorithm combined with `Bilat_7_1` filter presumably would give better reconstruction results when the noise levels are not known. Taking into account the number of sources the Bilateral filter with large windows size and smaller t_s value also would be a good choice. Summarizing our results we can state that in binary case of DT when we have noisy projection data it is suggested to combine other filters in the DART iterations in the way presented in Algorithm 1.

The results can be generalized to investigation of discrete phantoms where the objects consist more than one materials. It would also be interesting to see the influence of the filters when the objects have some property (e.g. when we reconstruct a circle like objects). It is still an open question that how the applied filters influence the necessary number of iterations during the DART reconstruction algorithm. We are going to perform our test framework on real projection data with the proposed DART variants.

Acknowledgments. This work was supported by the European Union and co-funded by the European Social Fund. Project title: "Telemedicine-focused research activities on the field of Mathematics, Informatics and Medical sciences." Project number: TÁMOP-4.2.2.A-11/1/KONV-2012-0073.

The author would like to thank Joost Batenburg and Christoph Schnör for the software phantoms. Another set of image phantoms were taken from the image database of the IAPR Technical Committee on Discrete Geometry (TC18). The author also would like to thank László Varga for the initial DART implementation.

References

1. Alpers, A., Poulsen, H.F., Knudsen, E., Herman, G.T.: A discrete tomography algorithm for improving the quality of three-dimensional X-ray diffraction grain maps. Journal of Applied Crystallography 39(4), 582–588 (2006)
2. Batenburg, K.J.: A network flow algorithm for reconstructing binary images from continuous x-rays. Journal of Mathematical Imaging and Vision 30(3), 231–248 (2008)

3. Batenburg, K.J., Sijbers, J.: Dart: A fast heuristic algebraic reconstruction algorithm for discrete tomography. In: IEEE International Conference on Image Processing, ICIP 2007, vol. 4, pp. IV–133–IV–136 (2007)
4. Batenburg, K., Sijbers, J.: Dart: A practical reconstruction algorithm for discrete tomography. IEEE Transactions on Image Processing 20(9), 2542–2553 (2011)
5. Chaudhury, K.N.: Acceleration of the shiftable $O(1)$ algorithm for bilateral filtering and non-local means. CoRR abs/1203.5128 (2012)
6. Chaudhury, K., Sage, D., Unser, M.: Fast $O(1)$ bilateral filtering using trigonometric range kernels. IEEE Transactions on Image Processing 20(12), 3376–3382 (2011)
7. Hantos, N., Balázs, P.: Image enhancement by median filters in algebraic reconstruction methods: An experimental study. In: Bebis, G., et al. (eds.) ISVC 2010, Part III. LNCS, vol. 6455, pp. 339–348. Springer, Heidelberg (2010)
8. Herman, G.T.: Fundamentals of Computerized Tomography: Image Reconstruction from Projections, 2nd edn. Springer Publishing Company, Incorporated (2009)
9. Herman, G.T., Kuba, A.: Advances in Discrete Tomography and Its Applications (Applied and Numerical Harmonic Analysis). Birkhauser (2007)
10. Kuba, A., Herman, G.T., Matej, S., Todd-Pokropek, A.: Medical applications of discrete tomography. In: Discrete Mathematical Problems with Medical Applications, DIMACS Workshop, DIMACS Center, Princeton, NJ, USA, December 8-10, pp. 195–208. AMS, American Mathematical Society, Providence (2000)
11. Maestre-Deusto, F., Scavello, G., Pizarro, J., Galindo, P.: Adart: An adaptive algebraic reconstruction algorithm for discrete tomography. IEEE Transactions on Image Processing 20(8), 2146–2152 (2011)
12. Nagy, A., Kuba, A.: Reconstruction of binary matrices from fan-beam projections. Acta Cybernetica 17(2), 359–385 (2005)
13. Pereira, L.F.A., Roelandts, T., Sijbers, J.: Inline 3d x-ray inspection of food using discrete tomography. In: InsideFood Symposium, Leuven, Belgium (2013)
14. Schüle, T., Schnörr, C., Weber, S., Hornegger, J.: Discrete tomography by convex-concave regularization and d.c. programming. Discr. Appl. Math. 151, 229–243 (2005)
15. Tomasi, C., Manduchi, R.: Bilateral filtering for gray and color images. In: Proceedings of the Sixth International Conference on Computer Vision, ICCV 1998, pp. 836–846. IEEE Computer Society, Washington, DC (1998)

Splitting of Overlapping Cells in Peripheral Blood Smear Images by Concavity Analysis

Feminna Sheeba[1], Robinson Thamburaj[2], Joy John Mammen[3], and Atulya K. Nagar[4]

[1] Department of Computer Science, Madras Christian College, Chennai
fsheeba@gmail.com
[2] Department of Mathematics, Madras Christian College, Chennai
robin.mcc@gmail.com
[3] Department of Transfusion Medicine & Immunohaematology, Christian Medical College, Vellore, India
joymammen@cmcvellore.ac.in
[4] Department of Mathematics and Computer Science Liverpool Hope University, Liverpool, L16 9JD, UK
nagara@hope.ac.uk

Abstract. The diagnosis of a patient's pathological condition, through the study of peripheral blood smear images, is a highly complicated process, the results of which require high levels of precision. In order to analyze the cells in the images individually, the cells can be segmented using appropriate automated segmentation techniques, thereby avoiding the cumbersome and error-prone existing manual methods. A marker controlled watershed transform, which was used in the previous study is an efficient technique to segment the cells and split overlapping cells in the image. However this technique fails to split the overlapping cells that do not have higher gradient values in the overlapping area. The proposed work aims to analyze the concavity of the overlapping cells and split the clumped Red Blood Cells (RBCs), as RBC segmentation is vital in diagnosing various pathological disorders and life-threatening diseases such as malaria. Splitting is done based on the number of dip points in the overlapping region using developed splitting algorithms. Successful splitting of overlapped RBCs help the count of the RBC's remain accurate during the search for possible pathological infections and disorders.

Keywords: Watershed transform, Overlapping cells, Concavity analysis, Curve smoothing, Dip points.

1   Introduction

Medical Imaging has become an essential component in various fields of biomedical research and clinical practice. It enables quantitative analysis of medical images, especially in blood smear images, thereby enhancing the ability to diagnose and treat various pathological disorders. Quantitative analysis can supplement clinicians with accurate results for the diagnosis of various diseases. A patient's pathological

R.P. Barneva, V.E. Brimkov, and J. Šlapal (Eds.): IWCIA 2014, LNCS 8466, pp. 238–249, 2014.
© Springer International Publishing Switzerland 2014

condition can be diagnosed by analyzing the various cells in his or her blood smear images. Typical blood smear images have a rich representation of different objects namely five types of White Blood Cells (WBCs), Red Blood Cells (RBCs) and Platelets. A Complete Blood Count (CBC) test gives important information about the kinds and numbers of these cells in the blood. A CBC test usually includes WBC and its types' count, Hematocrit, Hemoglobin. RBC Indices, Platelet Count and Mean Platelet Volume (MPV), involving counting of WBCs, RBCs and platelets. Another very important application is to detect malarial parasites in blood smear images. In such analyses, the RBCs need to be separately segmented, in order to check if the parasites reside in them. In order to get a higher success rate in RBC segmentation, all the clumped RBCs are required to be split to arrive at a correct cell count. Manual counting of cells by inspection of slides under the microscope is time consuming and prone to human error. With the increase in the number of blood samples to be examined by technicians, there is a need to automate the cell segmentation and counting process.

A simple segmentation technique to separate cells from the background of the image based on the gray values of the pixels is used, which does not touch upon splitting of overlapping cells [7, 12]. Another simple technique which does segmentation of WBCs in blood smear images using Tissue like P-systems was attempted which gave faster results [6]. A marker controlled watershed transform with morphological operations is used to segment the cells, for which the success rate is higher than the intensity based segmentation [3, 14]. However, this technique is only capable of handling touched or minor overlaps in the image. Distance transform is used in conjunction with the watershed algorithm so that many of the overlapping RBCs are split [12]. Separation and Counting of Blood cells are done using Geometrical Features and Distance Transform Watershed [10]. A technique in detecting touching gametocytes of malarial parasites is discussed in [4]. A new automatic algorithm to separate the touching and overlapping particles based on the intensity variations in the regions of touching and overlapping particles and geometric features of boundary curves has been developed . High Degree of overlapping cells can be split by focusing on rapidly detecting central point of the overlapping cells using the distance transform value [9]. Separating touching cells in hematoxylin stained breast TMA specimen is done by composing single-path voting followed by mean-shift clustering. The contour of each cell is obtained using a level set algorithm based on an interactive model [11]. A shape-based approach is proposed to do curve evolution for the segmentation of medical images containing known object types, which is able to handle multidimensional data, can deal with topological changes of the curve, is robust to noise and initial contour placements, and is computationally efficient. It avoids the need for point correspondences during the training phase of the algorithm [13]. A novel energy functional is proposed for segmenting neighboring organs, in which an extended maximum posteriori (MAP) shape estimation model is proposed using level set methods to recover organ shapes [17]. A multiphase graph cut framework is presented which simultaneously segments multiple overlapping objects. The advantage of this approach is that the segmentation energy is minimized directly without having to compute its gradient, which can be a cumbersome task and often relies on approximations [15]. A study to cluster the nuclei seen in confocal microscopy images was done using a clump-splitting algorithm to separate touching or overlapping nuclei allowing accurate detection of the number and size of the nuclei [2].

One recent common method used for segmenting overlapping or clumped cells is by doing their concavity analysis. A novel nonparametric concavity point analysis-based method for splitting clumps of convex objects in binary images is presented in [8]. The method is based on finding concavity point-pairs by using a variable- size rectangular window. Results obtained with images that have clumps of biological cells show that the method gives accurate results.

In peripheral blood smear images two or three RBCs overlap in various forms resulting into one or more concavities. In some of the overlapping cells, the gradient values in the overlapped region do not show remarkable difference compared to the other areas of the cell. Hence splitting them according to the concavity and convexity of the overlapped cells is more appropriate. The proposed system aims at splitting the overlapped RBCs of the blood smear images, which are left out after applying segmentation techniques like watershed algorithm to the images. Though watershed transform splits most of the overlapping cells, the very few ones that are not split are handled by our proposed system, which uses various splitting algorithms to split the overlapped cells. The paper is organized with methodology in Section 2, results and findings in Section 3 and conclusion in Section4 of the paper.

2   Methodology

2.1   Image Acquisition

The images for this study are obtained using a digital camera attached to a compound microscope. It uses a choice of 1.3 megapixel standard resolution or 2.9 megapixel high resolutions. The images acquired are 24 bit colored tiff images with a resolution size of 1280x1024 pixels. The software for the segmentation and watermarking was written in MATLAB R2011a.

2.2   Image Pre-processing

The input RGB images are first converted to binary images by using thresholding using Otsu's method. The images thus acquired may have poor illumination. Therefore, the contrast of the images is improved by equalizing the histogram of the image. Median filtering is used to remove noise from the images.

2.3   Initial Segmentation and Motivation to the Proposed Work

The authors' previous works include segmentation of various cells in peripheral blood smear images. Various segmentation techniques and feature extraction are used to classify the cells as RBCs, platelets or any one of the five categories of WBCs, leading to the calculation of their differential count. In addition, segmentation of gametocytes (fourth stage of malaria parasites) was done and an attempt was made to segment the ring structures (chromatin dots) of malarial parasites in the blood smear images. In all these segmentation processes, the RBCs were segmented first. In order to split the overlapped RBCs, watershed algorithm was applied. Most of the overlapped RBCs were split. However, it failed to split few overlapping cells with less intensity variations in the overlapping regions. This drawback motivated a need for the proposed work, where such overlapping RBCs are identified and appropriate algorithms are applied to split the overlapping cells.

2.4   Cell Splitting

The proposed study operates on binary sub-images with overlapping cells extracted from the initial segmentation. The overlapped cells form a single blob and based on their overlapping structure, the blob can be categorized either as an open or a closed object. The concave points of the overlapping cells (where the cells meet) are called as dip points. In the open object, the overlapping region is a single closed region, which is not the case in a closed object. When the segmented blood smear images were analyzed, it was found that the cells that were not split were of four types, which are shown in Figure 1 and 2. The blobs in figure 1(a), 1(b) and 1(c) are all closed objects with 1, 2 and 3 dip points respectively, whereas the blob in figure 2 is an open object with 4 dip points. The dip points and the overlapping structures are also shown in the figures.

Finding Dip Points
As a first step, the morphological operation close is applied to the object (overlapped cells) to smooth its boundary. A curve drawing algorithm suggested [16] involving tracing the boundary of overlapping objects, mathematical processing on tracing data and transformation to polar coordinates is used, resulting in a curve against rho and theta (rho is the distance from origin to the point and theta is the counterclockwise angle relative to the x-axis).

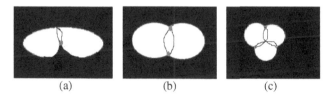

| (a) | (b) | (c) |

Fig. 1. Closed Objects

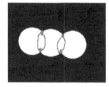

Fig. 2. Open Object

The second step is to identify the number of dip points in the overlapped cell. The convex hull of the object is first obtained. The Euclidean distance between the points in the convex hull of the object and the boundary points of the object are found by obtaining the distance transform matrix of the image consisting of the overlapping cell blob. The distance transform matrix consists of the distance of every pixel to its nearest non-zero value. When the image of the distance transform matrix is displayed the regions between the boundary of the object and its convex hull is displayed. This includes all the concave regions (overlapping region).The maximum distance value of each concave region is none other than the dip points.

The third step is to smooth the curve obtained in the first step using a moving average filter. Moving average filter smooth data by replacing each data point in the curve with the average of the neighboring data points defined within the span [16]. This process is equivalent to low pass filtering with the response of the smoothing given by the difference equation

$$y_s(i) = \frac{1}{2N+1}(y(i+N)+y(i+N-1)+...+y(i-N)) \tag{1}$$

Each time Smoothing is done, the minima points of the curve are obtained and the process is terminated when the minima points become equal to the number of dip points of the object.

Cell Splitting Methodology
Splitting methodology relies on the number of dip points. In the case of splitting the closed object with a single dip point as shown in Figure 3 (a), the dip point has to be joined with the boundary pixel in the opposite side of the object. To achieve this, the major axis of the object is first computed. A perpendicular connection is made from the dip point to the major axis, which is then extended to meet the border of the object in the same direction. The dip points are simply joined for objects with two dip points as shown in Figure 3(b) and for the ones with three dip points, the points are joined to the center point of the object as shown in Figure 3(c) [1].For objects with four dip points, orientation of the object decides the splitting. If the object lies almost parallel to x or y axis, the comparison of the dip points is done with the central point of the object to decide the joining of points. If the object lies diagonal to the x-y plane, comparison is done with the points in the major or minor axis of the object for splitting. These comparisons decide where the points lie. The points that lie in the same side (when compared with the central point) are joined resulting in splitting of three cells. This is shown in Figure 4 (a), 4(b), 4(c) and 4(d).

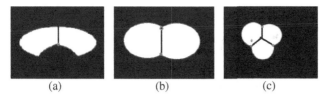

(a)   (b)   (c)

Fig. 3. Splitting Lines for Closed Objects

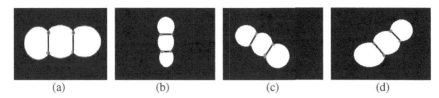

(a)   (b)   (c)   (d)

Fig. 4. Splitting Lines for Open Object

3   Results and Findings

3.1   Dataset

A dataset consists of sub-mages, which are in binary form, cropped from the segmented blood smear images. The previous work involves marker controlled watershed based segmentation technique applied to blood smear images in the RGB form. Three such RGB images are shown in Figure 5 and their corresponding watershed based segmented images are shown in Figure 6. It is seen from the figure, that the overlapped cells with clear concavity are split, whereas few of them having less concave regions are not split . Such cells are identified based on their geometric measures and are cropped to form sub-images, which is actually the dataset. The dataset consists of 50 such sub-images cropped from various segmented images.

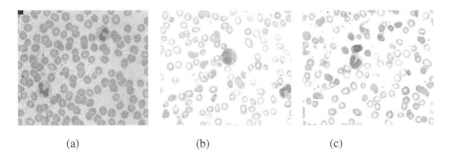

(a)   (b)   (c)

Fig. 5. Input RGB Images

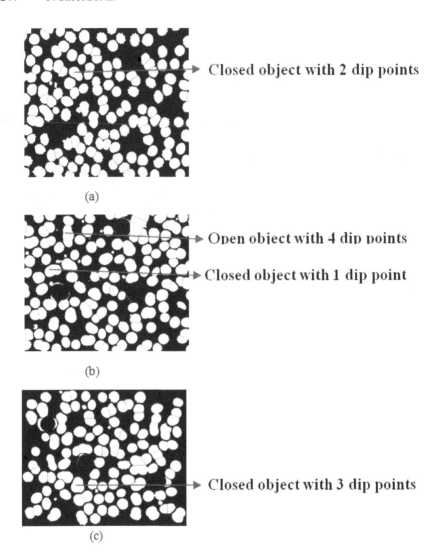

(a)

Closed object with 2 dip points

(b)

Open object with 4 dip points

Closed object with 1 dip point

(c)

Closed object with 3 dip points

Fig. 6. Segmented Binary Images with RBCs

Figures 6(a), 6(b) and 6(c) show all four types of overlapped cells that are not split by initial segmentation.

When Figure 5(a) is analyzed, it is seen that most of the overlapped cells are split by the watershed algorithm. Three cells that are not split are marked in Figure 7.

The three cells that are not split are identified using geometric measures and are separated as shown in Figure 8.

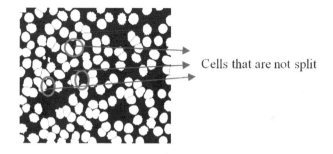

Fig. 7. Segmented Binary Image of Figure 6(a)

Fig. 8. Overlapped cells that are not split by Watershed algorithm

The sub-images containing the overlapped cell (seen as a single object), are extracted from the binary image. Their boundary pixels alone are extracted. The curve with the boundary pixels against rho and theta are drawn for each blobs. The dip points of the blobs are then found and displayed. Curve smoothing is done by applying moving object filter to each of the curves. The dip points are obtained using the method discussed in the Methodology section. Based on the number of dip points, the appropriate cell splitting method among the ones discussed was applied to the object. The step-by-step results are shown in Table 1. The sub-images after splitting the cells (as seen in the last row of Table 1) are then superimposed with image shown in Figure 8 and then with the original image, which is shown in Figure 9.

Fifty such sub-images were tested. Results of few more cells are shown in Figure 10. The first cell is a closed object with 2 dip points, next two are closed objects with 1 dip point , next three are closed objects with 2 dip points and the last two are closed object with 3 dip points and open object with 4 dip points.

Table 1. Step-by-step results for three objects

	Object 1	Object 2	Object 3
Sub-images			
Image Boundary			
Curve against rho and theta			
Smooth curve			
Concave regions			
Dip points			
Cell Splitting			

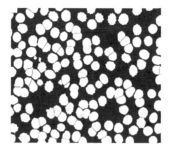

Fig. 9. Image with all overlapping cells split

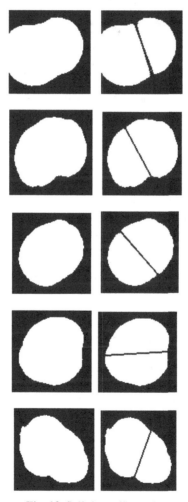

Fig. 10. Splitting cells results

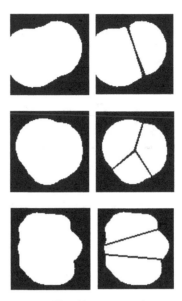

Fig. 10. (*continued*)

4   Conclusion

This study which involves splitting of overlapping RBCs in peripheral blood smear images improves the success rate of segmentation of cells in the images. The segmentation of overlapping cells left out by the watershed algorithm were captured and the spitting of such cells were done effectively, thereby increasing the accuracy in the diagnosis of pathological conditions and various diseases. Fifty overlapping cells were tested and the success rate was 75%. The work is still in progress as blood smear images have certain overlapping structures where the concavity regions are not clearly visible. Case by case study and evaluation has to be done to get accuracy closer to 100%.

Acknowledgement. The authors would like to thank the Centre for Applicable Mathematics and Systems Science (CAMSS) , Liverpool Hope University, UK for the support and funding towards the project work and the Department of Transfusion Medicine & Immunohaematology for providing them with sample images for their study. The authors would also like to acknowledge Mr. S. Kulasekaran for his contribution towards the development of the system that performs automatic segmentation of various cells in peripheral blood smear images.

References

1. Fan, J., Zhang, Y., Wang, R., Li, S.: A separating algorithm for overlapping cell images. Journal of Software Engineering and Applications 6(4), 179–183 (2013)
2. Farhan, M., Yli-Harja, O., Niemistö, A.: A novel method for splitting clumps of convex objects incorporating image intensity and using rectangular window-based concavity point-pair search. Pattern Recognition 46(3), 741–751 (2013)

3. Feminna, S., Robinson, T., Mammen, J.J., Thomas, H.M.T., Nagar, A. K. : White Blood Cell Segmentation and Watermarking. In: Proceedings of the IASTED International Symposia Imaging and Signal Processing in Healthcare and Technology, ISPHT 2011, Washington DC, USA (2011)

4. Feminna, S., Robinson, T., Mammen, J.J., Nagar, A.K.: Detection of plasmodium falciparum in peripheral blood smear images. In: Bansal, J.C., Singh, P., Deep, K., Pant, M., Nagar, A. (eds.) Proceedings of BICTA 2012. AISC, vol. 202, pp. 289–298. Springer, Heidelberg (2013)

5. Feminna, S., Robinson, T., Michael, J., Maqlin, P., Mammen, J.: Segmentation of sputum smear images for detection of tuberculosis bacilli. BMC Infectious Diseases 2012 12 (suppl. 1), O14 (2012)

6. Feminna, S., Robinson, T., Nagar, A.K., Mammen, J.J.: Segmentation of peripheral blood smear images using tissue-like P-Systems. IJNCR–BICTA 2011 Special Issue 3(1), 16–27 (2012)

7. Feminna, S., Thomas, H.M.T., Mammen, J.J.: Segmentation and reversible watermarking of peripheral blood smear images. In: Proceedings of the IEEE Conference on Bio Inspired Computing: Theories and Applications, vol. 2, pp. 1373–1376 (2010)

8. LaTorre, A., et al.: Segmentation of neuronal nuclei based on clump splitting and a two-step binarization of images. Expert Syst. Appl. 40(16), 6521–6530 (2013)

9. Nguyen, N.-T., Duong, A.-D., Vu, H.-Q.: Cell Splitting with High Degree of Overlapping in Peripheral Blood Smear. International Journal of Computer Theory and Engineering 3(3) (2011)

10. Prasad, A.S., Latha, K.S., Rao, S.K.: Separation and counting of blood cells using geometrical features and distance transformed watershed. International Journal of Engineering and Innovative Technology (IJEIT) 3(2) (2013)

11. Qi, X., et al.: Robust segmentation of overlapping cells in histopathology specimens using parallel seed detection and repulsive level set. IEEE Transactions on Biomedical Engineering 59(3), 754–765 (2012)

12. Sharif, J.M., et al.: Red blood cell segmentation using masking and watershed algorithm: A preliminary study. In: Proceedings of International Conference on Biomedical Engineering (ICoBE), Penang, Malaysia, pp. 258–262 (2012)

13. Tsai, A., et al.: A shape-based approach to the segmentation of medical imagery using level sets. IEEE Transactions on Medical Imaging 22(2), 137 (2003)

14. Tulsani, H., Saxena, S., Yadav, N.: Segmentation using morphological watershed transformation for counting blood cells. International Journal of Computer Applications & Information Technology 2(3) (2013)

15. Vu, N., Manjunath, B.S.: Shape prior segmentation of multiple objects with graph cuts. In: Proceedings of Computer Vision and Pattern Recognition, CVPR 2008 (2008), doi:10.1109/CVPR.2008.4587450, ISBN: 978-1-4244-2242-5

16. Yadollahi, M., Prochazka, A.: Segmentation for object detection, http://dsp.vscht.cz/konference_matlab/MATLAB11/prispevky/129_yadollahi.pdf (retrieved November 11, 2013)

17. Yan, P., Shen, W., Kassim, A.A., Shah, M.: Segmentation of neighboring organs in medical image with model competition. In: Duncan, J.S., Gerig, G. (eds.) MICCAI 2005. LNCS, vol. 3749, pp. 270–277. Springer, Heidelberg (2005)

Boundary Extraction for Imperfectly Segmented Nuclei in Breast Histopathology Images – A Convex Edge Grouping Approach

Maqlin Paramanandam[1], Robinson Thamburaj[1], Marie Theresa Manipadam[2], and Atulya K. Nagar[3]

[1] Department of Mathematics, Madras Christian College, Chennai, India
maqlinparamanandam@yahoo.com, robin.mcc@gmail.com
[2] Department of Pathology, Christian Medical College, Vellore, India
mtm2005@cmcvellore.ac.in
[3] Department of Mathematics and Computer Science, Liverpool Hope University,
Liverpool, L16 9JD, UK
nagara@hope.ac.uk

Abstract. The detection of cell nuclei plays a significant role in automated breast cancer grading and classification. Although many algorithms for nuclei detection are present in contemporary literature, there is a general arduousness in automatically segmenting nuclei which have an inhomogenous interior and weak boundaries revealed by uneven staining. Such nuclei are common in high grade breast cancer cells. This paper presents an automated boundary extraction methodology for detecting the broken or missing boundaries of imperfectly-segmented nuclei in breast histopathology images. The images are first segmented using K-means clustering method, to retrieve the prospective nuclei regions which may contain these imperfectly segmented nuclei. Following this, a boundary extraction methodology based on the grouping of approximately convex boundaries is used to uncover missing edges and connect the gaps inbetween them. The study is focused on patchy and open vesicular nuclei which are common in high grade breast cancers and which normally pose a challenge for automatic segmentation techniques. Using a sample size of a 100 images of nuclei for this evaluation, the proposed method yielded sensitivity and specificity rates of 90% and 93% with average Hausdorff distance measuring 59. In comparison, the same three factors achieved by employing color-based K-means clustering technique amounted to 49%, 92% and 323, whereas color deconvolution yielded 85%, 69% and 373 and intensity-based segmentation returned 14%, 97% and 351.

Keywords: Histopathology, Breast cancer, Nuclei segmentation, Boundary extraction, Edge grouping.

1   Introduction

Breast cancer is the most common cancer among women worldwide with a 1.67 million new cancer cases diagnosed in 2012 (25% of all cancers) [8]. Through several

R.P. Barneva, V.E. Brimkov, and J. Šlapal (Eds.): IWCIA 2014, LNCS 8466, pp. 250–261, 2014.

decades, microscopic examination of the biopsy tissue has remained the definitive standard for the diagnosis and grading of the disease. The breast cancer grading schemes are based principally on the pathologist's assessment of the three parameters in the biopsy specimen: acinar formation, nuclear size and shape irregularities (pleomorphism) and mitotic activity. Moreover in breast cancer, nuclear pleomorphism is graded and highly correlates with the aggressiveness of the disease and patient outcome. Studies conducted by researchers report breast cancer grading schemes to be highly subjective, leading to grading errors and have recommended replacing them with objective quantification techniques [12]. Nuclear pleomorphism is the most subjective element of histopathology grading and there is a considerable amount of observer variability in its grading. Normally breast pathology specialists tend to allocate higher grades than non-specialists [6]. Thus, automatic segmentation of cell nuclei and image-based objective quantification of nuclei pleomorphism would be invaluable for the breast cancer pathologists.

Accurate segmentation of cell nuclei has always been an important problem for automated breast cancer grading. Many nuclear segmentation algorithms proposed in the literature are précised in Section 3. This paper deals with segmentation of nuclei which do not have defined boundaries and appear patchy while revealing their contents. The work finds its significance from the fact that the histopathology tissue sections of nuclear pleomorphism is graded based on the size and shape of nuclei. Grade 2 and grade 3 breast cancers show cells which appear to have open vesicular nuclei with prominently visible nucleoli. These cells do not stain in an even way, but appear patchy and also seem larger in size when compared with normal nuclei. The outer rim of the cell appears to be thin and the staining may be concentrated there. The nucleolus of the cell is densely concentrated with stain and appears dark and circular. Perfect segmentation of such nuclei and analyzing their size and shape is invaluable in determining the nuclear pleomorphic grade of the tissue.

Most nuclei detection algorithms in the literature deal with isolated, overlapping and clumped nuclei without focusing on patchy vesicular nuclei which remain imperfectly segmented. Reference [4] deals with automated segmentation of nuclei in breast histopathology images and suggests that classification between benign and malignant images is not affected by imperfectly segmented nuclei. However, grading is a major prognostic marker in breast cancer and it requires proper determination of the shape and size, thereby helping to differentiate between normal nuclei and the cancerous ones [3].

This paper proposes a method to uncover the approximate boundaries of imperfectly segmented nuclei by detecting prospective nuclei regions in the image, selecting optimal boundary points from the prospective nuclei based on a convex grouping algorithm and finally extracting the boundary of the nuclei using the method of thresholding. The organization of this paper is as follows: Section 2 reviews the related works, Section 3 and Section 4 provide details on the proposed methodology and results of the work respectively and finally Section 5 concludes the paper.

2   Related Works

A detailed review of the histopathology image analysis works in the literature is given in [9]. Numerous methods have been proposed to segment the cell nuclei, extract their boundaries and also delineate overlapping and clumped nuclei.

The earliest studies conducted in the field of histopathology image segmentation have proposed methodologies such as thresholding, fuzzy c-means clustering, adaptive thresholding and watershed segmentation [10, 15 and 16]. In [16] an adaptive optimal thresholding (Otsu thresholding) segmentation and morphological opening and closing operations are used for completing the gaps and separating weakly connected nuclei in breast cancer histology images. The objective was to classify the segmented histological microstructures from high resolution histology and cytology images into inflammatory cells, lymphocytes, epithelial cells, cancer cells; and high nuclei density regions, but not to delineate nuclei boundaries accurately. The experimental results found 25% badly segmented nuclei, 4.5%–16.7% nuclei that remained in clumps, and missed 0.4%–1.5% nuclei in each image of H&E stained breast biopsies. Thresholding works only on images with uniform intensity. Watershed algorithms for nuclei detection discussed in [10] may result in a major drawback of over-segmentation caused by unwanted local minima.

In [15] nuclei of prostate and breast histopathology images were segmented using a Bayesian classifier, which incorporates low level information such as pixel color and texture, to create likelihood scenes of nuclei. The shape-based template matching algorithm was integrated along with the Bayesian classifier to detect nuclei from the likelihood scenes. Additionally incorporating other techniques for gland segmentation and nuclei feature analysis, the study showed an SVM classification accuracy of 80.52% for distinguishing high and low grade breast cancers compared to a manual classification accuracy of 99.33%.

An active contour model for overlapping resolution with watershed initialization (ACOReW) was discussed in [1] for segmenting nuclei and lymphocytes and resolving nuclei and lymphocyte overlapping. The method used shape based priors and region based active contours. Watershed segmentation results were used to initialize the active contour model. A quality evaluation of both nuclei and lymphocytes segmentation and resolving their overlapping were performed and compared with two other models called Geodesic Active Contour (GAC) and Rousson's shape-based model. The results showed that ACOReW outperformed the other two methods by accurately segmenting 92% of nuclei and 90% lymphocytes and also resolved 92.5% of overlapping resolutions. However, active contours are sensitive to initializations.

The Gradient in Polar Space model method proposed by [7] has been developed for extraction of nuclei in breast histopathology images. Nuclear regions are segmented by performing a gamma correction of the input image, followed by thresholding and morphological operations. The image is segmented into patches containing the cell nuclei. Every patch is transformed into a polar co-ordinate system. Finally, a median filter is applied for noise removal, and a biquadratic filtering is used to produce a gradient image from which nuclei boundaries are obtained. Size, Shape and textural

features extracted from segmented nuclei are used in nucleic-pleomorphism classification. The method claims that nuclear pleomorphism grading does not require segmentation of all nuclei in the histology image and can be achieved by segmenting only critical cell nuclei.

Basavanhally et al. proposed in [2] a method for tubule detection in breast histopathology images by incorporating domain knowledge, in which nuclei are segmented based on a color deconvolution method. Color deconvolution converts the image from the RGB color space to a new color space comprising hematoxylin H (i.e., purple), eosin E (i.e., pink), and background K (i.e., white) channels. On the other hand, the method does not perfectly segment nuclei which are not evenly stained and thus cannot be used for nuclear grading.

A marker controlled watershed method discussed in [17] uses fast radial symmetry to define the internal and back ground markers. The final segmentation results are approximated by ellipse fitting. Fast radial symmetry performs better than the regional minima markers. The nuclei segmentation accuracy was 79.2%.

Maqlin et al. Proposed in [14] a methodology to automatically detect tubules in breast histopathology images. In order to eliminate the falsely detected tubule regions the domain knowledge of the arrangement of nuclei around the tubule region was incorporated based on a direction and distance criteria. Here the segmentation of nuclei was achieved through level set evolution using a grid based initialization.

In [11] a Marked Point Processes (MPP) based model developed for multiple complex-shaped object extraction from images has been applied for nuclei extraction. MPP method takes into account the geometry of objects; and the information about their spatial properties by modeling inter-object relations in the scene showed by the image. The methodology was compared with two other state-of-the-art techniques namely, Gradient in Polar Space model (GiPS) and K. Mosaliganti's Level Set-based model (KMLS). MPP detected 1024 nuclei out of 1104 number of actual nuclei. The MPP model obtains the best F-measure of 0.7038 compared to KMLS method with 0.6292 and GiPS method with 0.5909.

3   Image Acquisition

The images for study comprise of histopathology images of breast biopsy specimens which are diagnosed with invasive ductal carcinoma by a pathologist at the Christian Medical College Hospital, Vellore, India. The images were acquired using a digital camera DFC280 attached to a compound microscope setup with a 40x objective. The images acquired are 24 bit colored JPEG images with a resolution size of 1280x1024 pixels. MATLAB R2009b is the computational tool used for developing the software. A total number of 16 images depicting invasive ductal carcinoma of the breast were used for the study. The images were manually classified by a pathologist and categorized under grade1, grade 2 and grade 3 nuclei pleomorphism. The vesicular cell nuclei which have inhomogeneous interior and weak boundaries are seen more in grade 2 and grade 3 images, which were used for evaluating the proposed methodology.

4   Methodology

The proposed method aims at detecting the boundaries of imperfectly segmented nuclei. The approach comprises of the following steps. 1) Detection of prospective nuclei regions 2) Selection of optimal points in the prospective nucleic-region for boundary extraction and 3) Extraction of definite nuclei boundary based on an edge grouping technique proposed in [18].

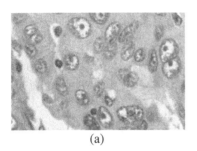 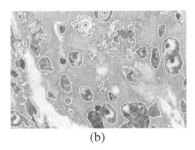

(a)   (b)

Fig. 1. (a). H&E stained Breast histopathology Image of Invasive ductal carcinoma, (b). Yellow contours showing prospective nuclei regions detected by K-means clustering

4.1   Detection of Prospective Nuclei Boundaries

The Histopathology Image, due to the H&E staining, shows three predominant colors: blue representing nuclei, pink representing stroma, and white representing lumen and the background. In order to detect prospective nuclei regions the image pixels are clustered into three color classes using the traditional K-means clustering algorithm [13]. Every pixel that belongs to a nuclear region falls within the blue cluster and hence represents the prospective nuclei regions in the Image. The K-means result of the blue cluster image, representing the prospective nuclei region, is converted into a binary image and each object boundary is extracted using the canny edge detection method [5]. Figure 1 shows the results of prospective nuclear-boundary detection.

4.2   Selection of Optimal Boundary Points

Each point in the boundary of the prospective nuclei is associated with a unique direction. The direction is specified based on the gradient vectors of the gray scale intensity image at that point.

The gradient vectors are obtained as follows: First the original image is converted into gray scale image followed by enhancement of the image using contrast-limited adaptive histogram equalization (CLAHE) [19]. CLAHE is used to limit the contrast, especially in homogeneous areas, to avoid reduce noise that might be present in the image.

The numerical gradient of the enhanced gray image defines the variation in intensity values along the horizontal and vertical direction. The gradient value defines the direction of each pixel in the image. Since the intensity variation between the nuclei

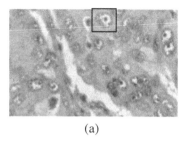 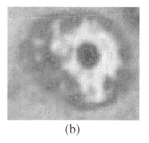

(a)   (b)

Fig. 2. (a). Direction of each boundary point of the prospective nuclei are plotted with green arrows, (b). A sub image showing single nucleus and the directions of its boundary points

and the tissue background were sharp in the histopathology images, the gradient values at the nuclei boundary points are at a peak and are found to have a direction pointing outward from the center of the nuclei. In order to direct the boundary points towards the centroids, the sign of gradient values are changed to negative values. Figure 2 illustrates a sub-image depicting single segmented prospective nuclei, for which the corresponding gradient vectors at its boundary points are plotted by green arrows.

The boundary of some prospective nuclei, especially the ones which become vesicular due to cancer, obtained from the K-means segmentation result, may contain spurious edges and gaps caused due to inhomogeneous staining. One such prospective nucleus is shown in a sub-image as in Figure 2 (b). The imperfect segmentation of such nuclei may affect the assessment and analysis of its shape, size and contents, which are significant for the automatic grading of cancer. Hence it is necessary to remove spurious edge pixels and complete the boundary gaps.

The boundary extraction can be done by selecting optimal boundary points from the prospective nuclei's boundaries. A boundary point is considered optimal, if it is oriented towards a reference point in the interior of the nuclei region at an angle less than $\pi/2$ (Figure 3). The reference point can be taken as the center of mass (centroids) of the set of boundary points for every prospective nuclei region. In simple terms, the optimal boundary points are the ones for which the angle between its direction and the line that connects it with the corresponding centroids is below $\pi/2$.

The selection of optimal boundary points is explained in Algorithm 1, where 'Opt' denotes the image containing optimal edge points, 'B', the set of prospective nuclei regions, 'b', the set of boundary points that belong to a prospective nuclei, 'g', the mass center of the set of boundary points 'b' of a prospective nuclei . 'P' represents each point in the prospective nuclei boundary set 'b'. 'd' is the direction of the boundary point p which is specified by the gradient value at the particular point in the gradient intensity image.

4.3   Extraction of Nuclei Boundary

The algorithm discussed in [18] finds the closed boundaries of a convex form, which optimally coincides with the optimal edges detected by edge detection algorithms.

Algorithm 1. Selection of optimal boundary points

```
opt  = 0;
for  ∀ b ∈ B
  g = mass    center    of  b
  for ∀ p ∈ b
        n = g − p
        α = cos⁻¹ n̂ · d̂

        if  α < π/2
              opt(p)    = 1
        end  if
  end   for
end   for
```

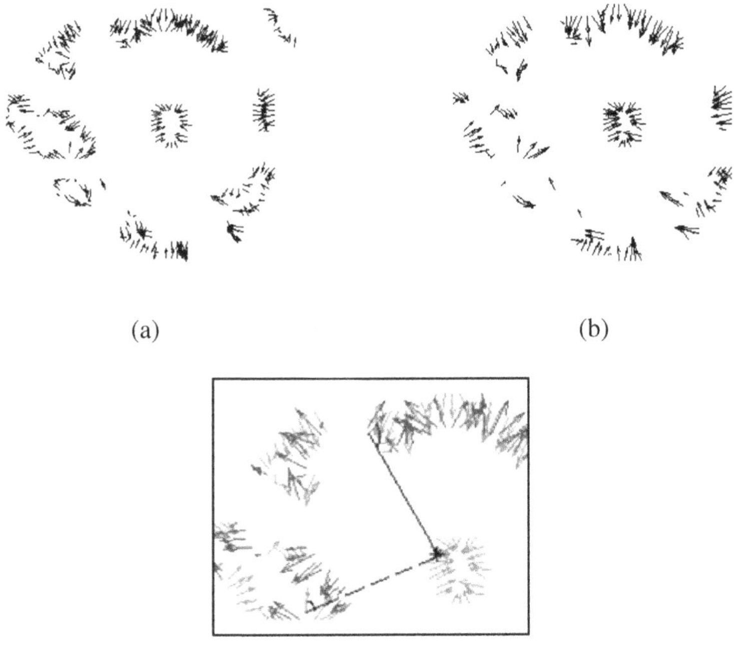

(a)   (b)

(c)

Fig. 3. (a) Black arrows showing the gradient directions of boundary points and red dot showing the mass center of the boundary points (b) optimal boundary points oriented towards the mass center. (c) A sub image of a, with green arrows showing the directions of optimal boundary points (oriented towards the mass center), red arrows showing directions of other boundary points , dashed lines connecting an optimal boundary point to mass center, solid lines connecting a non optimal boundary point to the mass center.

Since nuclei are objects confined within convex boundaries, the same technique is used in this paper for finding the boundary of prospective nuclei. It is assumed that the boundary coincides with the optimal boundary points that have been detected in the previous step. The two components of the boundary extraction algorithm are: Building an accumulator image and Thresholding the accumulator Image to get the nuclei boundary.

Building an Accumulator Image

This step exploits the property that at least one line passes through each boundary point of a convex set (which are the nuclei's boundary points in this work), such that the convex set lies entirely within the two closed half planes. More of this concept is explained in [18]. In our case, the convex set is the nucleic region. The optimal boundary points detected in previous step is the approximate convex boundary. The aim is to detect all points in the region of interest that falls within the nucleic region and thereby determining the exact nuclear boundary.

Region of interest: In order to select the region of interest, a circular region is considered, whose center is the mass center of the corresponding nuclei boundary points and which has an empirically determined radius.

Accumulator Image: An image of height and width equal to the original image and initialized with zeros at every pixel location is taken as the accumulator image. The accumulator image is built by a voting algorithm. For each point within the region of interest, if the point lies inside the half plane defined by an optimal boundary point in that region, then the value of the corresponding pixel location in the accumulator image is incremented by 1. This is repeated for every point within the region of interest.

Given a point R within the ROI, the dot product of R-P and P-G is computed in order to determine whether a ROI point lies within the half plane of a boundary point P. Here G is the mass center of all the optimal boundary points within the region. If the result is a positive value, then it is concluded that the ROI point is within the half plane of the boundary point. Every pixel value of the accumulator image is incremented by the number of boundary points within whose half planes it lies. The accumulator image thus built is converted into an 8 bit gray intensity image.

Thresholding the Accumulator Image

The gray image is enhanced using CLAHE, as shown in Figure 4, and then converted to a binary image using Otsu's thresholding. The boundary of the nuclei is thus obtained.

   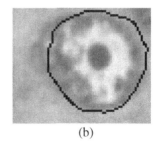

(a)   (b)

Fig. 4. (a) Accumulator Image, (b) Detected Nuclei boundary by thresholding

5   Results and Discussion

The proposed boundary extraction algorithm was evaluated for a sample size of 100 sub-images containing open vesicular or patchy nuclei. These images were cropped from 5 different images of breast histopathology sections which were diagnosed for grade 2 and grade 3 Invasive Ductal Carcinoma. In order to evaluate the performance of the proposed method, the results are both qualitatively and quantitatively compared with the Color-based K-means clustering, Color deconvolution, Intensity based segmentation. The qualitative and quantitative results are laid out in the following subsections.

5.1   Qualitative Results

Each image in column (a) of Figure 5 shows a sub-image of an open vesicular or patchy nucleus. Columns (b), (c), (d) and (d) depict the segmentation results obtained by color based K-means clustering, color deconvolution, intensity based segmentation

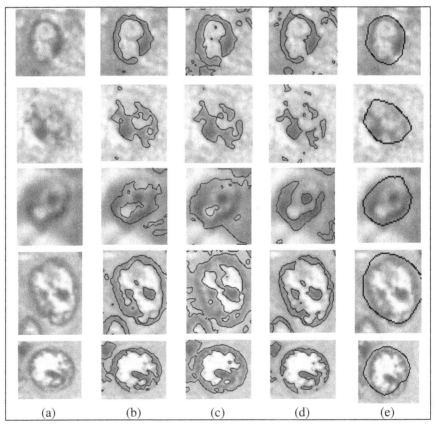

| (a) | (b) | (c) | (d) | (e) |

Fig. 5. (a) Cropped Images of nuclei, (b) Results of color based color based K-means clustering, (c) Results of color based color deconvolution, (d) Results of intensity based segmentation and (e) Boundary detection by the proposed method

and the proposed boundary extraction method, in order of appearance. The results reveal that the proposed method performed better in comparison to other methods, by extracting the boundary of the entire nuclei, in spite of them being unevenly stained with faint edges and inhomogeneous interiors.

5.2   Quantitative Results

Three measures namely Sensitivity, Specificity and the Hausdorff distance are used to evaluate the proposed method in comparison with color based K-means clustering, color deconvolution and intensity based segmentation. A manual delineation of the nuclei was performed for all the hundred images and the results were used as the gold standard for evaluation.

Sensitivity is the true positive rate that indicates the effectiveness of the algorithm's performance in identifying nuclei pixels.

$$\text{Sensitivity} = \frac{\text{TruePositive}}{\text{Truepositive} + \text{TrueNegative}} \tag{1}$$

True Positive rate is the portion of nuclei pixels which are both within the manually segmented nuclei region and the automatically extracted nuclei boundary.

Specificity is the false positive rate that indicates the effectiveness of the algorithm's performance in identifying pixels which are not nuclei.

$$\text{Specificity} = \frac{\text{FalsePositive}}{\text{FalsePositive} + \text{FalseNegative}} \tag{2}$$

False Positive rate is the portion of non-nuclei pixels which are both outside the manually segmented nuclei region and the automatically extracted nuclei boundary.

The Hausdorff distance is used to measure the accuracy of the boundary points obtained by the proposed algorithm. This is primarily a measurement of the distance computed between the two given contours. The smaller the distance value, the more similar the contours are to each other.

$$h(A, B) = \max_{a \in A} \{ \min_{b \in B} \{ d(a, b) \} \} \tag{3}$$

$$H(A, B) = \max \{ h(A, B), h(B, A) \} \tag{4}$$

$h(A,B)$ is the Hausdorff distance between two sets of contour points A and B , 'a' is a point in the set A (manually segmented nuclei boundary points) , 'b' is a point in the set B (automatically segmented nuclei boundary points), and $d(a, b)$ is the distance between the point 'a' and point 'b'. The smaller the Hausdorff distance the more accurate the segmentation algorithm performs in detecting the boundary of the nuclei. The results of the quantitative evaluation are shown in Table 1. Upon quantitative evaluation the proposed method has shown to obtain higher sensitivity and specificity rates and strikingly smaller Hausdorff distance compared to other methods and thus proves to be a better method for segmenting open vesicular or patchy nuclei.

Table 1. The average and standard deviation of Sensitivity, Specificity and Hausdorff distance of the segmentation algorithms for 100 images

	Sensitivity	Specificity	Hausdorff distance
Color-based K-means Clustering	0.49±0.194	0.92±0.8	323±146
Color Deconvolution	0.85±0.1	0.69±0.15	373±137
Intensity based Segmentation	0.14±0.02	0.97±0.02	351±142
Proposed method – Boundary Extraction based on convex edge grouping	0.90±0.03	0.93±0.06	59±26

6   Conclusion

The proposed boundary extraction algorithm, uses the approximately convex boundaries obtained from the segmentation results in order to find missing boundaries. Unlike many segmentation approaches in contemporary literature, this method does not concentrate on cell overlapping and clumped nuclei regions. The contribution mainly focuses on nuclei which are vesicular, patchy, unevenly stained rather than dense ones. The analysis of these nuclei is quite significant for breast cancer grading. The study can be extended to deal with detection of cell overlapping and delineation of individual nuclei from clumped cells.

Acknowledgement. This work was supported by the Centre for Applicable Mathematics and Systems Science (CAMSS, Liverpool Hope University, UK. The authors would like to thank the Department of Pathology, Christian Medical College Hospital, Vellore, India, for providing the sample images for the study.

References

1. Ali, S., Madabhushi, A.: Active contour for overlap resolution using watershed based initialization (ACOReW): Applications to histopathology. In: IEEE International Symposium on Biomedical Imaging: Nano to Macro, pp. 614–617 (2011)
2. Basavanhally, A., Yu, E., Xu, J., Ganesan, S., Feldman, M., Tomaszewski, J., Madabhushi, A.: Incorporating domain knowledge for tubule detection in breast histopathology using O'Callaghan neighborhoods. In: SPIE Medical Imaging Medical Imaging: Computer-Aided Diagnosis, vol. 7963, p. 796310 (2011)
3. Boucheron, L.E., Manjunath, B.S., Harvey, N.R.: Use of imperfectly segmented nuclei in the classification of histopathology images of breast cancer. In: IEEE ICASSP, pp. 666–669 (2010)
4. Bussolati, G.: Proper detection of the nuclear shape: ways and significance. Rom. J. Morphol. Embryol. 249(4), 435–439 (2008)
5. Canny, J.: A computational approach to edge detection. IEEE Trans. Pattern Anal. Mach. Intell. 8(6), 679–698 (1986)
6. Dabbs, D.J.: Breast Pathology. Saunders Print Book (2012)

7. Dalle, J.R., Li, H., Huang, C.H., Leow, W.K., Racoceanu, D., Putti, T.C.: Nuclear pleomorphism scoring by selective cell nuclei detection. In: IEEE Workshop on Applications of Computer Vision (2009)

8. GLOBOCAN 2012, International Agency for Research on Cancer, World Health Organization (2012), http://globocan.iarc.fr/Pages/fact_sheets_cancer.aspx

9. Gurcan, M.N., Boucheron, L.E., Can, A., Madabhushi, A., Rajpoot, N.M., Yener, B.: Histopathological image analysis: A review. IEEE Reviews in Biomedical Engineering 2, 147–171 (2009)

10. Karvelis, P.S., Fotiadis, D.I., Georgiou, I., Syrrou, M.: A watershed based segmentation method for multispectral chromosome images classification. Proc. IEEE Eng. Med. Biol. Soc. 1, 3009–3012 (2006)

11. Kulikova, M.S., Veillard, A., Roux, L., Racoceanu, D.: Nuclei extraction from histopathological images using a marked process approach. In: Proc. SPIE Medical Imaging (2012)

12. Ladekarl, M.: Objective malignancy grading: A review emphasizing unbiased stereology applied to breast tumors. APMIS Suppl. 79, 1–34 (1998)

13. Lucchese, L., Mitra, S.K.: Unsupervised segmentation of color images based on K-means clustering in the chromaticity plane. In: Proc. IEEE Workshop on Content-based Access of Images and Video Libraries, pp. 74–78 (1999)

14. Maqlin, P., Robinson, T., Mammen, J.J., Nagar, A.K.: Automatic detection of tubules in breast histopathological images. In: Bansal, J.C., Singh, P., Deep, K., Pant, M., Nagar, A. (eds.) Proc. of Seventh International Conference on Bio-Inspired Computing: Theories and Applications (BIC-TA). AISC, vol. 202, pp. 311–321. Springer, Heidelberg (2013)

15. Naik, S., Doyle, S., Madabhushi, A., Tomaszeweski, J., Feldman, M.: Automated gland segmentation and Gleason grading of prostate histology by integrating low-, high-level and domain specific information. In: MIAAB Workshop (2007)

16. Petushi, S., Garcia, F.U., Haber, M.M., Katsinis, C., Tozeren, A.: Large-scale computations on histology images reveal grade-differentiating parameters for breast cancer. BMC Med. Imag. 6, 14 (2006)

17. Veta, M., Huisman, A., Viergever, M.A., van Diest, P.J., Pluim, W.: Marker-controlled watershed segmentation of nuclei in H&E stained breast cancer biopsy images. In: Symposium on Biomedical Imaging (ISBI), pp. 618–621. IEEE (2011)

18. Zingman, I.: Novel algorithm for extraction of an object with an approximately convex boundary. In: Proc: 7th IASTED International Conference on Signal Processing, Pattern Recognition and Applications (2010)

19. Zuiderveld, K.: Contrast limited adaptive histogram equalization. In: Graphics Gems IV, pp. 474–485 (1994)

A Noise Adaptive Fuzzy Equalization Method with Variable Neighborhood for Processing of High Dynamic Range Images in Solar Corona Research

Miloslav Druckmüller and Hana Druckmüllerová

Institute of Mathematics, Faculty of Mechanical Engineering,
Brno University of Technology
Technická 2, 616 69 Brno, Czech Republic
druckmuller@fme.vutbr.cz, ydruck00@stud.fme.vutbr.cz

Abstract. This paper presents a generalization of the Noise Adaptive Fuzzy Equalization Method (NAFE) which was developed for visualization of high dynamic range (HDR) images produced by the Atmospheric Imaging Assembly (AIA) aboard the Solar Dynamics Observatory (SDO) spacecraft launched by NASA in 2010. This generalization widens the usability of the NAFE method to HDR images with extreme brightness gradient and with significantly different parts. This type of images is typical for imaging of solar corona by means of white-light coronagraphs and during total solar eclipses however the method may have much wider field of usage even outside the solar research.

Keywords: Solar corona, Visualization of high dynamic range images, Noise adaptivefuzzy equalization method, Image processing.

1 Introduction

Visualization of HDR images represents today a very important branch of research in image processing. The reason is the continuously increasing gap between sixteen or even more bits per pixel digitization of images and eight bits per pixel dynamic range used by displays. It has no sense to decrease this gap because of the ability of human vision not to distinguish more brightness levels than about 256 on the screen on the contrary to distinguishing much more levels watching real scenes and adapting its sensitivity and other properties in various parts of the scene. In the principle, it is not possible to create a universal method for visualization of HDR images because of different and often opposing requirements for the output. HDR images typical for solar corona imaging have several very specific properties. These images are generally of very low (local) contrast, but they contain bright features with extreme contrast. These contrasty features are magnetic loops in active regions, which are typical for extreme ultraviolet images (SDO AIA, Proba 2, STEREO, SOHO EIT) and prominences in white-light images. The contrast of the features is so high that no monotonous pixel

R.P. Barneva, V.E. Brimkov, and J. Šlapal (Eds.): IWCIA 2014, LNCS 8466, pp. 262–271, 2014.
© Springer International Publishing Switzerland 2014

Fig. 1. Examples of typical HDR images of solar corona. Left solar corona in extreme ultraviolet at wavelength 171 Å obtained by SDO AIA spacecraft ($\gamma = 2.2$), right white-light corona acquired during total solar eclipse ($\gamma = 4.5$).

value transform $y = t(x)$ (where x is the input and y the output pixel value) is able to produce an image suitable for computer display or for printing (typically 8 bit per pixel, i.e. 256 brightness levels), namely an image which would show all features with a contrast sufficient for human vision. That is why a compromise is usually made, whereby the processed images contain saturated parts in order to expose the dark, low contrast, features. Figure 1 shows two examples of images of the solar corona which have the described properties.

Various methods have been developed for processing images of the solar corona outside the solar disk, which is dominated by a steep gradient of brightness. The first method that enables one to visualize both large- and small-scale structures and that respects the properties of human vision was the proposed by Druckmüller et al. in [2]. A fast method that was proposed by Morgan and al. in [7], the Normalizing-Radial-Graded Filter (NRGF), and it is nowadays used as a standard tool both for images from coronagraph and images of the solar corona taken during total solar eclipses in spectral lines. This method was later extended by Druckmüllerová et al. [3] in Fourier-Normalizing-Radial-Graded Filter, which makes it possible to visualize finer details compared to the NRGF. These advances in image processing techniques led to many observations that would be impossible to be achieved with alder methods and also with less powerful computers of that times. Among that observations are the studies of the fine structures visible during the total solar eclipse of 2006 (Pasachoff et al. [8]), especially the first observation of rapid changes in the polar region (Pasachoff et al. [9]). More recently, high-resolution observations of the solar corona both in white light and in spectral lines brought the team of Habbal to new results in the knowledge about prominences [4] and coronal mass ejections [5].

A classical method for visualization of fine low-contrast structures is based on attenuating low spatial frequencies by means of techniques based on the Fourier transform (or other orthogonal transforms) or convolution [10]. If we use the convolution of the original image A with kernel C for that purpose, the resulting image B may be written in the form

$$B = A * C = A * (D - L) = A - A * L = A - M \tag{1}$$

where $*$ denotes convolution, D denotes the Dirac kernel (i.e. a kernel with the property $A * D = A$), L denotes the kernel of a low-pass filter. The image M is usually called the unsharp mask. That is why the method is often called unsharp masking. As the highest contrast is typically in the low spatial frequencies, the resulting image B has usually a significantly lower dynamic range than the original image A, and is therefore more suited for visualization. Unfortunately, this is not the case for HDR images of the solar corona, where the high-contrast features are typically in the high spatial frequencies. That is why this classical method is not usable, because makes the situation even worse.

There is another class of methods, which is based on adaptive histogram modification, in particular, the adaptive histogram equalization [10]. These methods solve the problem of high dynamic range and extreme contrast very well, but they suffer from three critical problems (see Fig.2).

1. Appearance of artifacts caused by the shape of the neighborhood (usually a square) from which the histogram is computed. These artifacts may influence coronal structures significantly, leading to false interpretation.
2. Extreme amplification of noise in the low contrast parts of the image, resulting in the loss of faint details.
3. Loss of contrast on boundaries between areas with significantly different brightness.

Fig. 2. An HDR image of the solar corona during a total solar eclipse (left) processed by the adaptive histogram equalization with square neighborhood (right). The image illustrates all drawbacks of adaptive histogram equalization method - artifact caused by using of square neighborhood, loss of contrast near lunar edge (edge effect) and significant noise in areas with very low contrast.

The Noise Adaptive Fuzzy Equalization method (NAFE), which is described in [1], is inspired by both the adaptive histogram equalization and unsharp masking, and is in some sense a combination of both methods. The method solves problems 1 and 2. Problem 3 is solved by the Noise Adaptive Fuzzy Equalization Method with Variable Neighbourhood (NAFE-VN) presented in this paper.

2   The NAFE-VN Method

Let A denote the original input image. In this image the pixel values have a linear dependence on the intensity of the coronal emission. Let B denote the resulting image. This image is a linear combination of two images: $T_\gamma(A)$ and $E_{N,\sigma}(A)$, such that

$$B = (1 - w)T_\gamma(A) + wE_{N,\sigma}(A). \tag{2}$$

$T_\gamma(A)$ denotes the so-called gamma transform of image A, which is produced by the pixel value transform

$$y = b_0 + (b_1 - b_0)\left(\frac{x - a_0}{a_1 - a_0}\right)^{\gamma^{-1}}. \tag{3}$$

The constant a_0 is the input value representing 0 emission intensity. The constant a_1 is the maximum valid input value. Constants b_0, b_1 are minimum and maximum output values, respectively. The standard γ value for a PC computer display is $\gamma = 2.2$. Lower values for γ give darker images, and higher values give brighter images. Image $T_\gamma(A)$ in Equation (2) represents the 'unmodified' image upon which only a nonlinear pixel value transform was applied. $E_{N,\sigma}(A)$ is an image created by the Noise Adaptive Fuzzy Equalization (NAFE-VN) method, to be described in what follows. The constant w will be called the NAFE-VN weight. Typical values of w lie in the interval $\langle 0.05, 0.3\rangle$. Constant $w = 0$ gives an image B without any enhancement. The specification of a value for w enables one to control the enhancement of the structures (see Fig. 3).

The NAFE-VN method produces a pixel value transform which is not constant for the whole image like a gamma transform, it is not one function for the whole image. It is different for every pixel and is dependent on the neighborhood of the pixel. An inspiration for this method was the histogram equalization method, which is based on the idea that the optimal brightness level distribution in the image is the uniform one, i.e. all pixel values in the resulting image are used with the same probability.

Let us assume, for simplicity, that pixel values in image A are realizations of a continuous random variable V with a distribution function $F(x)$. The pixel value transform $y = mF(x)$ creates values of pixels in image B which are realizations of a random variable U with the uniform distribution on the interval $\langle 0, m\rangle$ [6]. In reality, the random variable V is discrete, that is why U is only an approximation of the uniform distribution. The images to be processed have typically several thousand discrete pixel values. Therefore the approximation is very good and the difference between a continuous and a discrete variable is negligible. Since we do not know the random variable V, only its realization (the image), the

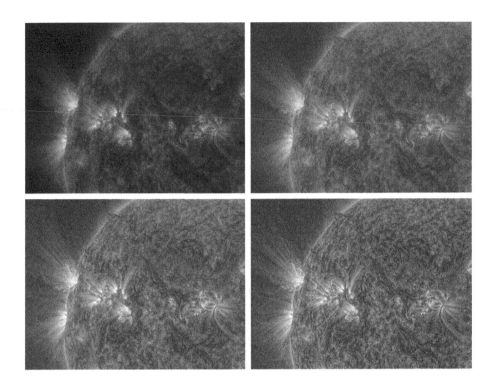

Fig. 3. From top-left to bottom-right: Original SDO AIA image at wavelength 171 Å(γ = 2.2)and NAFE-VN processed images with NAFE-VN weight w = 0.1, $w = 0.2$, $w = 0.3$

distribution function $F(x)$ must be estimated with the normalized cumulative histogram of the image.

The adaptive histogram equalization [10] is based on the idea that the cumulative histogram is not computed for image A as a whole, but only for the processed pixel neighborhood. In this case the pixel value transform function in not a compromise derived from the whole image but it is optimally set according to the pixel neighborhood properties. The typical size of the neighborhood is 10^4 or more pixels, which makes the algorithm extremely time consuming. That is why the square neighborhood is usually used, which enables the use of a recurrent algorithm for histogram computation.

The result of the adaptive histogram equalization is highly dependent on the size and shape of the neighborhood and the resulting image may contain severe artifacts. The reason may be explained by means of an analogy with convolution - Equation (1). Let us denote the square neighborhood by $N_{i,j}^n$, where $[i, j]$ are coordinates of the pixel, and the odd integer n is the width and height of the neighborhood in pixels. $N_{i,j}^n$ plays an analogous role of the convolution kernel L in which all elements are identical. Such a kernel is in principle a very bad low pass filter with a complicated Fourier spectrum which is direction dependent.

That is why kernels with better properties are used - for example a Gaussian kernel L. A significant improvement of the adaptive equalization may be achieved by using the fuzzy neighborhood $\widetilde{L}_{i,j}^n$ instead of the neighborhood $N_{i,j}^n$.

Let us denote by $L = [l_{k,l}]$ the matrix of size $n \times n$ (in analogy with L in Equation (1)), whose elements are in the interval $\langle 0, 1 \rangle$ and

$$k, l \in \left\{ \frac{1-n}{2}, \ldots, 0, \ldots, \frac{n-1}{2} \right\}. \tag{4}$$

The fuzzy neighborhood $\widetilde{L}_{i,j}^n$ is a fuzzy set [11] with support $N_{i,j}^n$ and membership function $\mu_{i,j}^n : N_{i,j}^n \to \langle 0, 1 \rangle$ where the membership grade of pixel $a_{i+k,j+l}$ to the fuzzy neighborhood is defined as

$$\mu_{i,j}^n(i + k, j + l) = l_{k,l}. \tag{5}$$

Now let us define the fuzzy histogram of $\widetilde{L}_{i,j}^n$ as

$$h_{i,j}^n(x) = \sum_{k=\frac{1-n}{2}}^{\frac{n-1}{2}} \sum_{l=\frac{1-n}{2}}^{\frac{n-1}{2}} l_{k,l}\, \delta_{x, a_{i+k,j+l}} \tag{6}$$

where δ denotes the Kronecker delta. Then we define the cumulative fuzzy histogram

$$H_{i,j}^n(x) = \sum_{m=a_0}^{x} h_{i,j}^n(m) \tag{7}$$

and the normalized cumulative fuzzy histogram

$$C_{i,j}^n(x) = H_{i,j}^n(a_1)^{-1} H_{i,j}^n(x), \tag{8}$$

where a_0, a_1 are minimal and maximal pixel values in $N_{i,j}^n$. Finally, we define the fuzzy equalizing function

$$f_{i,j}^n(x) = b_0 + (b_1 - b_0) C_{i,j}^n(x), \tag{9}$$

where b_0, b_1 are minimum and maximum output pixel values. This function is, unlike Equation (3), different for every pixel $a_{i,j}$, and the output pixel $q_{i,j}$ is computed according to formula

$$q_{i,j} = f_{i,j}^n(a_{i,j}). \tag{10}$$

The use of the fuzzy equalizing function solves the Problem 1 of the classical histogram equalization.

Next, we will solve Problem 3. The fuzzy histogram (6) is computed from all pixels in the processed pixel neighborhood regardless of its values. If the position of the processed pixel is near the border of areas A_1 and A_2 with significantly different pixel values, the neighborhood contains pixels from both areas. The fuzzy equalizing function (9) in this case is a compromise of optimal pixel value transforms for A_1 and A_2, which results in a significant decrease of local contrast. The solution is to use for fuzzy histogram computing only pixels that belong to that of area A_1 or A_2 which the processed pixel belongs to. Therefore, we replace (6) with formula

$$h_{i,j}^{n,\epsilon}(x) = \sum_{k=\frac{1-n}{2}}^{\frac{n-1}{2}} \sum_{l=\frac{1-n}{2}}^{\frac{n-1}{2}} l_{k,l}\, \delta_{x,a_{i+k,j+l}}\, \Delta_{x,a_{i+k,j+l}}^{\epsilon} \,, \tag{11}$$

where

$$\Delta_{x,a_{i+k,j+l}}^{\epsilon} = \begin{cases} 1 & \text{if } |x - a_{i+k,j+l}| < \epsilon \\ 0 & \text{else.} \end{cases} \tag{12}$$

The value of ϵ must be found experimentally. Too small values of ϵ cause image fragmentation into small areas with very high contrast features whose borders do not represent relevant boundaries in the image. Too high values of ϵ will result in identical or nearly identical images those obtained when applying Equation (6). Let us denote the corresponding normalized cumulative fuzzy histogram by $C_{i,j}^{n,\epsilon}(x)$ and the corresponding fuzzy equalizing function by $f_{i,j}^{n,\epsilon}(x)$. The use of this function solves both Problems 1 and 3.

However, serious problem 2. still persists, namely the extreme amplification of additive noise in areas with very low contrast, which results in the loss of faint low-contrast details. If the neighborhood $N_{i,j}^{n,\epsilon}$ contains noise only, the full dynamic range of the output image is used for noise display. The solution is to add more (artificial) additive noise to the input image. Of course not to the image itself, to the image which is used for histogram computing instead. This will decrease the contrast of the (natural) original additive noise in the resulting image. The presented solution is not very practical, because it is required to work with two different input images and the additive noise must be added to one of them by a noise generator.

Since the original noise in the image and the artificial noise are stochastically independent, an equivalent solution is to compute the convolution of $C_{i,j}^{n,\epsilon}(x)$ with the probability density function of the noise which we would be added to the image [6]. Let us suppose that the added noise has a Gaussian distribution with mean value $\mu = 0$, standard deviation σ. Let us denote its probability density function by $G_\sigma(x)$. Then let us define the noise adaptive fuzzy equalizing function

$$g_{i,j}^{n,\epsilon}(x) = b_0 + (b_1 - b_0)C_{i,j}^{n,\epsilon}(x) * G_\sigma(x). \tag{13}$$

This function is used for creating image $E_{N,\sigma}(A)$ in Equation (2). The convolution of the normalized cumulative fuzzy histogram with a Gaussian kernel has a significant influence only in those pixel neighborhoods that are dominated by noise, i.e. in which the image has very low contrast. On the other hand, the influence is negligible in the contrasty parts of the image. The optimal value of σ is typically in the interval $\langle 2\sigma_A, 12\sigma_A \rangle$ where σ_A is the standard deviation of the additive noise in the input image A. The higher is the value of σ, the higher is the amount of noise in the low contrast parts of the image. It is not correct to

Fig. 4. Example of noise reduction, SDO AIA 171 Å NAFE-VN processed images, with $w = 0.2$, and $\sigma = 10, 20, 30, 40, 50$, from left to right, respectively

suppress the noise too much, because the low contrast details will be lost. The noise must be clearly visible in the processed image but its intensity must not mask the visibility of fine details (see Figure 4). By applying the noise adaptive fuzzy equalizing function (13), all Problems 1, 2, and 3 are solved (see Figure 5).

3   Implementation and Parameters Setting

The main drawback of the NAFE-VN method is that it is extremely time consuming, therefore parallel implementation was necessary. Parallel implementation of this method was done in Borland Delphi (Pascal). SDO AIA images of $4\,096 \times 4\,096$ pixels size were used for testing. Processing of one image takes approximately 6 min on a PC with Intel Core i7-3820 working on 3.6 GHz (8 processor kernels), the processing of $1\,024 \times 768$ pixels crop of such an image takes approximately 20 s. It should be mentioned that the processing time depends not only on image size but also on the image structure. Low contrast parts are processed faster than contrasty ones.

The following processing parameters were used. Matrix $L = [l_{k,l}]$ of the size $n \times n$, $n = 129$, was defined in the form

$$l_{k,l} = d \sum_{m=1}^{12} c_m G_{\sigma_m}(k, l), \quad \sigma_m = 2^{\frac{m}{2}} \tag{14}$$

where $d, c_1, c_2, \ldots, c_{12}$ are constants and

Fig. 5. White-light solar corona HDR image obtained during a total solar eclipse (left) processed with NAFE-VN (right). Compare this image with Figure 2. The NAFE-VN processed image illustrates that none of mentioned problems is present.

$$G_{\sigma_m}(k,l) = \left(2\pi\sqrt{\sigma_m}\right)^{-1} e^{-0.5(k^2+l^2)\sigma_m^{-2}}.$$ (15)

The form of matrix L as a linear combination of Gaussian kernels G_{σ_m} was chosen because the constants $c_1, c_2, \ldots, c_{12}$ give enough variability necessary for fine tuning the properties of the fuzzy neighborhood. The constant

$$d = 2\pi \left(\sum_{m=1}^{12} c_m \sigma_m^{-1/2}\right)^{-1}$$ (16)

ensures that maximum element $l_{0,0} = 1$. After extensive testing on SDO AIA images, it was found that the optimum setting is very near to $c_m = 1$ for all $m = 1, 2, \ldots, 12$.

4 Conclusion

The NAFE-VN method described in this paper is a powerful tool for qualitative studies of the fine structures in HDR images. The method yields artifact-free images in which the local contrast, edge effect on significant boundaries and noise may be well controlled. The main disadvantage of the method is the processing time. Parallel implementation of this method is appropriate. The method was tested on SDO AIA images and white-light images of the solar corona obtained during total solar eclipses, but it can be used in many other applications, where HDR images are present.

Acknowledgement. Solar Dynamic Observatory (SDO) is a NASA project. This work was supported by Grant Agency of Brno University of Technology, project FSI-S-11-3. Publication of the results was financially supported by the project Popularization of BUT R&D results and support systematic collaboration with Czech students CZ.1.07/2.3.00/35.0004.

References

1. Druckmüller, M.: A noise adaptive fuzzy equalization method for processing solar extreme ultraviolet images. The Astrophysical Journal Supplement Series 207, 25–29 (2013)
2. Druckmüller, M., Rušin, V., Minarovjech, M.: A new numerical method of total solar eclipse photography processing. Contributions of the Astronomical Observatory Skalnate Pleso 36, 131–148 (2006)
3. Druckmüllerová, H., Morgan, H., Habbal, S.R.: Enhancing coronal structures with the Fourier Normalizing-Radial-Graded Filter. The Astrophysical Journal 737, 88–97 (2011)
4. Habbal, S.R., Druckmüller, M., Morgan, H., et al.: Total solar eclipse observations of hot prominence shrouds. The Astrophysical Journal 719, 1362–1369 (2010)
5. Habbal, S.R., Druckmüller, M., Morgan, H., et al.: Thermodynamics of the solar corona and evolution of the solar magnetic field as inferred from the total solar eclipse observations of 2010 July 11. The Astrophysical Journal 734, 120–137 (2011)
6. Hogg, R.V., McKean, J.W., Craig, A.T.: Introduction to Mathematical Statistics, 6th edn. Pearson Prentice Hall (2005)
7. Morgan, H., Habbal, S.R., Woo, R.: The depiction of coronal structure in white-light images. Solar Physics 236, 263–272 (2006)
8. Pasachoff, J.M., Rušin, V., Druckmüller, M., Saniga, M.: Fine structures in the white-light solar corona at the 2006 eclipse. The Astrophysical Journal 665, 824–829 (2007)
9. Pasachoff, J.M., Rušin, V., Druckmüller, M., et al.: Polar plume brightening during the 2006 March 29 total eclipse. The Astrophysical Journal 682, 638–643 (2008)
10. Pratt, W.K.: Digital image processing: PIKS inside, 3rd edn. Wiley, New York (2001)
11. Zadeh, L.A.: Fuzzy sets. Information and Control 8, 338–353 (1965)

Author Index